1 PATCHOULI
2 VETIVER
3 CISTUS
4 SANDALWOOD
5 LAVENDER
6 DAMASK ROSE
7 BERGAMOT
T0091972

# IN SEARCH OF PERFUMES

## A LIFETIME JOURNEY TO THE SOURCE OF NATURE'S SCENTS

DOMINIQUE ROQUES

*Translated from the French by*
*Stephanie Smee*

HARPERVIA
*An Imprint of* HarperCollins*Publishers*

*For my father,*

*who showed me the way to the trees*

Maps and line drawings © Emily Faccini.
Black and white photographs courtesy of the author.
Photograph on page 142 courtesy of Agence-odds & Authentic Products.



HarperCollins books may be purchased for educational, business, or sales promotional use. For information, please email the Special Markets Department at SPsales@harpercollins.com.

Originally published as *Cueilleur d'essences: Aux sources des parfums du monde* in France in 2021 by Éditions Grasset & Fasquelle.

FIRST HARPERVIA EDITION PUBLISHED IN 2023

*Design adapted by Ad Librum from the UK edition designed by Libanus Press Ltd.*

Library of Congress Cataloging-in-Publication Data is available upon request.

ISBN 978-0-06-329795-1

23 24 25 26 27 LBC 5 4 3 2 1

"May you stop at Phoenician trading stations
to buy fine things,
mother of pearl and coral, amber and ebony,
sensual perfume of every kind—
as many sensual perfumes as you can"

C. P. Cavafy
"Ithaka"

# CONTENTS

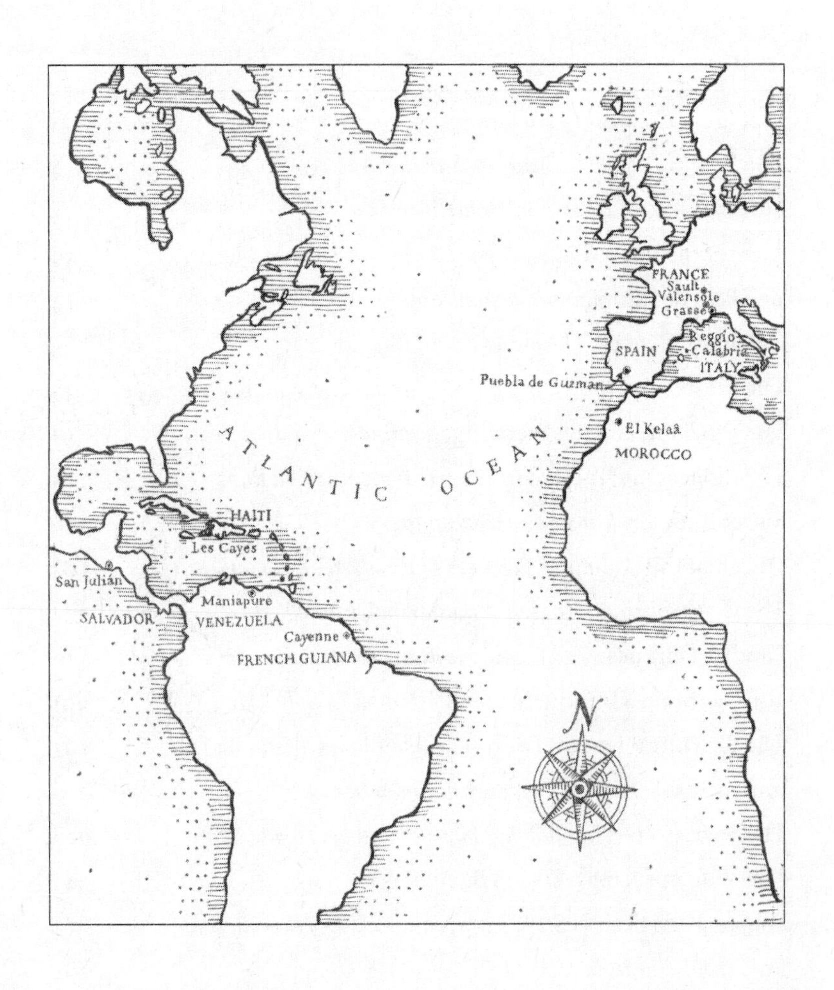
FRANCE
Sault
Valensole
Grasse
Reggio Calabria
SPAIN
ITALY
Puebla de Guzman
El Kelaâ
MOROCCO
ATLANTIC OCEAN
HAITI
Les Cayes
San Julián
SALVADOR
Maniapure
VENEZUELA
Cayenne
FRENCH GUIANA
N

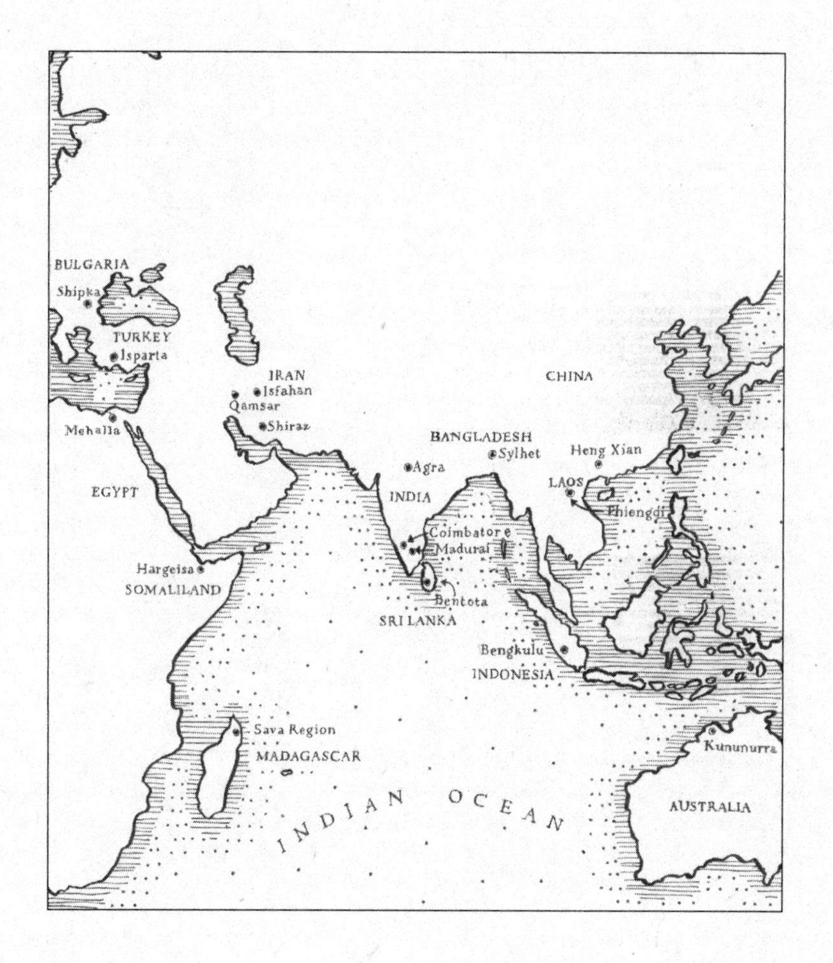
BULGARIA
Shipka
TURKEY
Isparta
IRAN
Isfahan
Qamsar
Shiraz
Mehalla
EGYPT
Hargeisa
SOMALILAND
CHINA
BANGLADESH
Sylhet
Agra
INDIA
Heng Xian
LAOS
Phiengdi
Coimbatore
Madurai
Bentota
SRI LANKA
Bengkulu
INDONESIA
Sava Region
MADAGASCAR
INDIAN OCEAN
Kununurra
AUSTRALIA

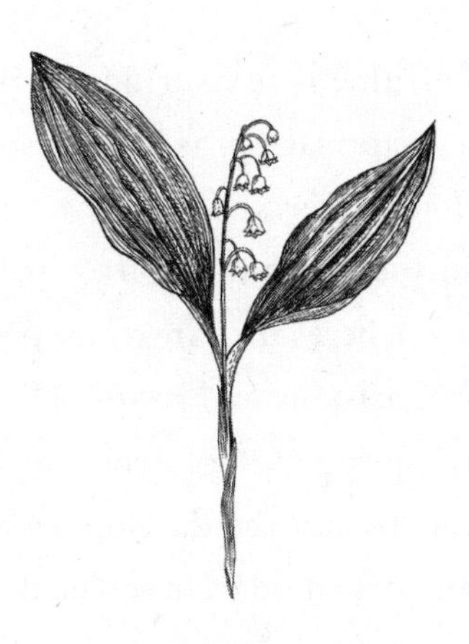

# PROLOGUE

## *The world's harvesters*

Perfumes are at once familiar to us, yet mysterious. They summon up fragments from the recesses of our olfactory memory, snatches of childhood recollections, as vivid as they are distant. There is no escaping it. Everybody carries with them through life a waft of lilac, a country lane lined with broom, the scent of loved ones. I clearly remember a moment I experienced as a child in the woods. It was May, and there was such a profusion of lily of the valley beneath the great oaks of the Rambouillet forest that the air was heavy with fragrance. I was spellbound, troubled by this scent that conjured up images of my mother wearing that sumptuous perfume, Diorissimo, a homage to those little white bells. A sense of intimate familiarity from the interplay of smell and memory, coupled with the mysteriously evocative power wielded by a composition when a flacon of scent is opened. Perfume is reassuring at first, as it reminds us of who we are, then captivating, as its own story is revealed.

*Here are fruits, flowers, leaves and branches*... Verlaine's familiar verse is a lyrical introduction to nature's own vast catalogue of scents. Let me expand: here, also, are roots, peels, woods, lichens, seeds, buds, berries, balsams and resins. The plant world, in all its guises, is the wellspring of the essences and extracts that have given us perfumery. Before the development in the nineteenth century of the chemistry of odorant molecules, natural products had for three millennia been the sole materials used in perfumes. While they have come to symbolize luxury in the industry, perfumers remain resolutely enamored with these natural scents. They bring a richness and complexity to their creations; indeed, some are perfumes in their own right.

Before evaporating on our skin, the formulae take only a moment to tell the tangled stories of their numerous component parts. Tales of laboratories, in the case of chemical ingredients, and of flowers, spices and resins for natural products. Distilled or extracted, these plants are transformed into essential oils, absolutes or resinoids,* to become part of a perfume's composition, taking their place alongside synthetic molecules. The olfactory depth of natural ingredients renders them indispensable in fine fragrances, and they always feature prominently in the marketing materials of perfume houses.

Essences tell their own story, the result of a coincidence of regions, landscapes, soils and climates, and the work of people who may have deep roots in that terroir or who may simply be

* The glossary at the end of the book contains definitions of technical terms.

passing through. The fragrance industry has always needed—and will continue to need—woodcutters for its aromatic woods such as cedar, oud and sandalwood. It still needs people to gather the plants that grow wild: the juniper berries, cistus branches and tonka beans. Collectors of saps and resins tapping trees for frankincense, benzoin and Peru balsam. Growers of flowers, leaves and roots, such as rose and jasmine, vetiver and patchouli. People to press citrus fruits such as bergamot and lemons. Carriers and merchants, successors to the caravaneers of Arabia and the mariners who linked India to Mediterranean lands. And, finally, distillers: the masters of rose water, alchemists of essences dating back to the seventeenth century—extractors, and chemists in the modern age. A community both disparate and dispersed, harvesting in deserts and forests, laboring with hoes and tractors, conducting deals that may be clandestine or transparent, unaware perhaps of their products' final destination, or, then again, receiving visits to their fields from renowned perfumers and representatives of the most distinguished houses.

It is a diverse world that has resulted, unwittingly, in a grand, historical community, creating a tapestry whose warp and weft tell us tales of lavender, rose and frankincense. Enigmatic itineraries, shifting origins, traditions that have been safeguarded, misplaced, lost and rediscovered; these are the stories of the making of perfumes, their creators all nourishing our enduring passion for nature's scents. When a Malagasy farmer pollinates a flower on her vanilla plant, there is a form of magic at play. It is an action she must repeat a thousand times over in order for the pods to form, to ripen, before they can be harvested, extracted and ultimately transformed into the delicious aroma of a little vial of vanilla absolute.

This book is an account of three decades of wandering, on the hunt for the source of the world's scents. Neither a chemist nor a botanist, I went to work in the perfume industry after studying business administration, thereby indulging an abiding interest in trees and plants. It was a journey prompted by appreciation and curiosity, a journey that developed into a passion, and for the last thirty years I have devoted myself to searching for, discovering, purchasing and, from time to time, producing essences for the fragrance industry. Whether it be in fields of roses or patchouli, in the forests of Venezuela or the villages of Laos, I have been initiated into a universe of scent by the people of these perfumed lands. They have taught me to listen to the stories told by the essences and extracts when their flacons are opened, and I have become what these days might be described as a "sourceror" or sourcing agent.

I work for a company that specializes in creating fragrances and flavors, and my role is to ensure that our perfumers are supplied with essences or extracts from more than one hundred and fifty natural raw materials from about fifty countries. My job involves securing consistent volume and quality, but I am also constantly on the hunt for new ingredients to extend the perfumers' "palettes." I am the first link in the organizational chain of this industry that stretches from fields of flowers to the flacons in a perfumery. The final protagonists in this story are the perfume houses themselves, and with the launch of every new product, perfumers from the various creative companies—the famous "noses," creators of complex and confidential formulae known as "juices"—are pitted against each other. The community of perfumers, a florilegium of talent and forceful personalities, is always conjuring up new scents for the most prestigious labels and that is where my experience in the field comes in.

My travels in perfume first started when I was working for a family-owned company based in the heart of the Landes forest. I became involved in setting up distillation and extraction facilities in countries where some of the major aromatic products are grown. A pioneer in its field in the 1980s, this company had pursued a policy of establishing facilities at source to produce natural extracts. Be it in Spain, Morocco, Bulgaria, Turkey or Madagascar, this involved installing equipment, organizing the cultivation and harvesting of crops, and managing production teams. I discovered places steeped in history, processes based on traditional know-how in danger of disappearing, and complex webs of human relationships.

For the last ten years, I have worked as a sourcing agent for a Swiss company, also family-owned, and one of the major global businesses involved in manufacturing fragrances and flavors. In order to supply and expand the catalogue of natural ingredients available to our perfumers, over the years I have helped develop with producers from around the world a network of partnerships that has allowed me to rub shoulders with people in every sector of the perfume industry. My passion for fragrances has been enhanced by every one of these encounters.

The geography of our products brings a sourcing agent face to face with a mosaic of social, economic and political realities. I have worked with numerous communities, many of them remote, vulnerable to the risk of cyclones and droughts, abandoned even by their own governments. Very early on, I became aware of our industry's role and responsibilities in the fate and future of these populations. It is a responsibility that both motivates me and guides the way I approach my work.

The inspiration for this book sprang from a recent trip, when I found myself standing next to a frankincense tree in the mountains of Somaliland. The collector who was accompanying me had just made an incision in the trunk, causing small milky drops to start to bead. Along with the intoxicating smell of the emerging frankincense, the wind carried with it a feeling of witnessing, at that very moment, the continuation of an extraordinary story, the story of the harvesting of nature's perfumes, a story that had persisted uninterrupted for more than three thousand years. Breathing in the scent of the fresh resin brought back memories from years earlier of my experience in the cistus, or rock rose, fields of Andalusia. I realized with a start that, from labdanum cistus to frankincense, I had had the good fortune over the previous thirty years to meet the heirs to this story that had endured for thousands of years. I knew then what I wanted to write: an account of perfume's source materials throughout the ages, the story of the lives of those who continue to devote themselves to their production, an account which would reflect the scope of their knowledge and traditions, the beauty of the places where they produce their scents, and the fragility of their future. Each stage of this story is different and unique, but there is a common thread: every element is part of a process to produce fragrances that move us profoundly. What better illustration of this than a fact I discovered in Bulgaria's Valley of Roses? In order to produce one kilogram of rose oil, one million flowers must be picked, by hand.

This book is my homage to the harvesters of the world.

# THE TEARS OF CHRIST

*Cistus in Andalusia*

Rounding a bend one April afternoon in the Andalusian countryside of El Andévalo, I found myself dazzled by the spectacle of fields of flowering rock rose, also known as labdanum cistus. It was a harbinger of the delights that would follow when I discovered the fragrance produced by that soil and the people who harvest that perfumed plant. At the end of the 1980s, the landscape of cistus-covered hills began as you left the town of Huelva, from the first village you came to in the hinterland. The road climbed up between plantations of eucalyptus, then snaked through vast expanses of twiggy shrubs whose leaves shone in the sunlight. Another village and then the great evergreen oaks appeared, standing alone, majestic sentinels at the gate, casting their shadow over that sun-scorched cistus country.

My fatigue after the 1,300-kilometer drive from France heightened my sensitivity to this new landscape. I was in Andalusia to establish from scratch a distillation and extraction plant. It was my first immersion in the world of fragrance and everything was new: the job, this region, its smells, its traditions. I spoke some basic Spanish, but I was going to have to make myself understood, recruit a team, set up a small factory and procure supplies for it. The aim was to cover the cistus extract requirements for a large perfume manufacturing group, a challenge that might prove out of my reach.

On this spring day, the hills were spangled with fat white flakes, as if an unlikely storm had dusted the fields in snow before making way for the Andalusian sunshine. Cistus flower between March and April. Their white flowers resemble poppies, as delicate as tissue paper, and last only two or three days. I set off on foot into this landscape of dense branches, where the vegetation was thick and difficult to penetrate. The cistus reached waist-high, sometimes higher, and the leaves on their branches were already gleaming. As soon as the blossoms appear, the plant starts secreting a resin, the famous labdanum gum that will coat the new growth through the summer, protecting it from the heat. A delectable scent floated across the hillside, not yet as intense as it would become in July but already addictive. The scent of the gum is as powerful as it is sticky. It has a warm odor, almost animalic, with an astonishing intensity. Cistus extract is ubiquitous in perfumes, its amber notes indispensable in oriental accords. Labdanum is an essential component of the mythical formula for Guerlain's Mitsouko, which in 1919 launched the revolutionary "chypre" accords, a previously unheard-of marriage of floral

*A mother and daughter boiling cistus twigs,*
*starting the labdanum gum production*

notes with exotic, spicy scents. The flowers themselves have no smell, they simply look superb: five white petals, a heart of yellow stamens and, at the base of each petal, a carmine-colored spot known by the Andalusians as "the tears of Christ." The cistus flower is part of their cultural heritage.

It was here in Andalusia that I first encountered this industry, a whole world, that would engage and captivate me for all these years, leading me on to wherever aromatic plants are grown. The perfumes offered up to us by these plants originate far from perfumeries, in the natural world where time moves slowly. They emerge from the earth, are then harvested, transformed and transported, the product of converging stories mysteriously combined in order, ultimately, to become an elixir in a flacon. To open a perfume is to experience that quick flash of surprise, of pleasure, the brief moment when the extract has a chance to reveal itself to us. Was it the scent of the resin, the fragile beauty of the flowers, or the feeling of having entered into the realm of a unique plant? Regardless, on that afternoon in spring, I was launched on an aromatic and emotional journey from which I have never truly returned.

I still remember Josefa. There in the middle of a field one summer's afternoon, in the cistus-covered hills, this Roma mother was directing her daughters in the cooking of the labdanum gum. In the furnace of the Andalusian summer, under her straw hat, fork in hand, she busied herself with the drums in which the cistus branches were boiling away, her tracksuit stained with resin, her face blackened by smoke. Seeing me arrive, she called out in a loud voice, "Well now, Mr. Frenchman, how's your Spanish coming along?" We spoke of the unbearable heat generated by the combination of sun and fire, and then of the gum she was preparing for me. "For the miserable amount you're paying us for it, you ought to be showering us in perfumes from Paris! When's the Chanel coming out?" she asked me, laughing. In her mind, perfume evoked a world of luxury she could only ever imagine. In a few words, her exclamation captured the great distance between those boiling the cistus and the flacons of perfume themselves, two peculiarly different extremes of an otherwise shared story.

Cistus labdanum, or *Cistus ladanifer*, is a shrub that grows wild along the shores of the Mediterranean, from Lebanon to Morocco. Where the soil is acidic, it quickly takes over uncultivated land. In some areas, conditions are so favorable that it grows in great swathes of hundreds, if not thousands, of hectares. Formerly found on Cyprus and Crete, these days it is found in Spain, particularly in south-western Andalusia where fields of it spread toward Portugal, growing among the cork oaks.

Labdanum gum was one of the very first aromatic materials to be used for its scent. References to it appear on Mesopotamian tablets dating back as early as 1700 B.C. The ancient Egyptians were familiar with the gum and would burn it, together with frankincense and myrrh. Stories of its harvest in ancient times create a picturesque image. The flocks of goats that used to roam the fields on Crete and Cyprus would return in the evenings, their fleece impregnated with resin, which the shepherds would collect with a carding comb in order to make a paste to burn. As time went on, the gum would be collected with rakes fitted with leather straps that would be used to beat the branches. The gum would then be scraped from the straps with a knife. Returning from my excursions into the fields, with gum stuck to my clothes, I liked to imagine the Cypriot shepherds around the fire in the evenings, scraping the gum into balls, precursors to our incense sticks.

As I would discover with Josefa and the Roma, the processing of gum is still a laborious job involving soda and sulfuric acid. A specialty of the Salamanca region in the pre-war period, the industry shifted after the war toward southern Spain, with its great expanses of cistus in Extremadura and Andalusia, before ultimately coming to a halt by the ocean at the very tip of the peninsula.

El Andévalo lies in the hinterland of Huelva province, not far from Portugal. Historically a mining region, it was a source of tin and silver in ancient times and, from the nineteenth century, of copper and iron pyrites. But in the 1980s, the mines closed at Rio Tinto, and soon all that remained was the river that flows red as a result of the iron ore, and its name, appropriated by the world's largest mining company. Left behind is a landscape with a metallic earth that every now and then seems to tremble, and a culture of hard-boiled, rural miners. A land of strong traditions, a population deeply rooted in the land. Mining, hunting, horses, flamenco dancing and singing; white villages with cobbled streets where every year the population gathers in pilgrimage, creating in those who live there a true sense of community.

Puebla de Guzmán is the village we had selected for the facility. A crossroad in the region's hinterland, Puebla encompasses all of the area's component parts: mining life with its immense open-pit excavation, where the only sound still to be heard is the echo of cawing crows, a nascent Iberian pig-breeding industry which produces the famous *pata negra* ham, horses that are trained in Puebla as well as in Cádiz and Jerez and which strut their stuff on the weekends, the hunting of partridges fond of nesting in the cistus-covered hills, bars where you can breakfast on toast drizzled with olive oil. And then there are the fiestas where every generation knows how to dance the *sevillanas* and where there will always be a singer and somebody with a guitar ready to break into a *cante flamenco*, that reflection of the Andalusian soul.

I had hired a team of a dozen workers from the village. Over the moon to have found work after the closure of the mine, these men were heirs to a strong working-class culture, trade unionism included. Andalusians, steeped in tradition, endearing fellows. A year after the first earthworks, the factory had started production, and bundles of gleaming branches sat piled in the sun outside the main workshop, waiting to be crushed and distilled. The scent of cistus from the factory drifted deep into the surrounding countryside and passers-by would eye the facility from the road, not a little proud to see their village having made the change from mining to perfume. The extraction of iron pyrites was being replaced by cistus extraction; there was clearly something exceptional about their soil.

The man responsible for opening my eyes to the region was a pig breeder and farm manager called Juan Lorenzo, who ensured our factory's supply of branches and gum. A fine example of pure Andévalo stock, farmer, breeder and hunter, Juan

Lorenzo was a taciturn man with an enduring love of the land. He knew everything there was to know about cistus. He was an impressive representative of his region with his cap, his forthright gaze, and the hands of a man who has spent his days working the fields, and we enjoyed some fine moments together once I was able to make sense of his Andalusian dialect. He lived in an extraordinary farmhouse wedged into the hills beneath holm oaks, a white building that stood alone at the end of the road leading from the mine, where he kept a few horses and a hundred or so pigs of the finest origin. The quality of any future ham is measured by the number of days the animals have spent roaming freely among the oaks. At the end of the 1980s, *jamón de bellota* was not as famous as it is nowadays. A local delicacy, little known outside the region, it left any visitor in thrall to the unique taste imparted by the acorns to its flesh and fat.

Bit by bit, Juan Lorenzo brought me up to speed with community affairs. Puebla has always been surrounded by a vast area of cistus. When left unpruned, the plants grow to over two meters tall with very hard, woody stems which have traditionally been used by bakers to fuel their bread ovens. The region has, for decades, been an example of a balanced agropastoral farming model. The cistus grow beneath holm oaks, whose acorns help fatten up the pigs in winter. When the cistus get too old, they are pulled out and the earth is plowed for the sowing of wheat or oats. The following year, the cistus take over the fallow field once more and in two or three years' time there is a new uniform cover of young plants. This cycle of land management works well for the region's larger landholdings, farms that are several thousand hectares in size and owned by wealthy individuals or hunting associations based in Seville or Madrid. The region is known for its game and the cistus has an important role to play in that regard. Its thickets provide shelter for partridges and hare, and wild boar are never far from the holm oaks' acorns.

Another thing I learned from Juan Lorenzo is that the extraction of labdanum gum has long been performed by the Roma people. Living in Andalusia for long enough now that we can probably say forever, these people settled in the area after their forebears had set off from northern India and Pakistan on a journey spanning several centuries, a story that is not well known and has had more than its share of tragedy. There were villages in that part of Andalusia with significant Roma populations who would harvest the cistus and produce the gum. Several years later, I would come across other Roma villages when I went to Bulgaria to plant roses. There, on the other side of Europe, the Bulgarian Romany are as important to the cultivation of roses as the Roma of Andévalo are to the production of gum. There exists a striking symmetry between the presence and roles of these Roma communities at the two extremities of the continent. Having settled in these places, the families have one foot in the local culture and the other in their own way of life. Taciturn and not given to ostentation, the Roma tend to keep to themselves. In answer to my questions about their past, they would reply with cheerful banter and laughter. How long have they been in the business of producing gum? Who really knows, their fathers were doing it before them. It turns out that, in this region, the job of "gum distiller" is quite a recent occupation, dating no further back than the 1950s. For a long time, cistus was collected on the banks of the Tagus before the work spread south to where vast expanses of the plants grow, unmatched anywhere else in Europe. Wherever they might be, picking cistus in the west, roses in the east, marginal and marginalized, the Romany people are harvesters. The role played by these communities in the creation of these legendary products is not often acknowledged. But is this something which troubles them?

In his job as manager, it fell to Juan Lorenzo to determine which plots were ready to be cut according to the growing cycles of his various properties. It was always an early start to the day with him, small cups of very strong coffee and some bread with olive oil and local cheese at the café. Invariably, some Roma would join him, and lengthy negotiations would ensue. Not in Spanish but in Andalusian, that dialect which involves swallowing certain syllables in order to emphasize one's point! We were going out to the fields of some enormous properties, assessing the quality of the branches, determining access, quantities. He had his strategies for securing cistus without having to pay for it, in return for hours spent plowing fields that would be sown with wheat. He knew all the clans in the Roma villages in the area, indispensable connections, because gum is a family affair. He would introduce me to them, and my position as foreign manager of the business seemed to earn me their approval. We would then put in an order for an amount of gum according to the number of barrels each family would produce over the summer.

Deep in the countryside, at the end of a track that runs for several kilometers, one or two Roma families have been granted access to the fields of an estate and have set up a summer work station to produce labdanum gum. The site requires easy access to water, ideally to one of the small streams that do not dry up over the summer and which can be identified by the wild oleanders growing along their banks. The seasonal work station consists of a dozen old 200-liter oilcans next to which a ditch must be dug to catch the liquid run-off at the end of the process.

The morning is spent preparing bundles of branches, before the summer heat makes the work unbearable. Cutting cistus may seem easy enough. But to do it quickly and effectively, without exhausting oneself, is an art. The tool used is a broad sickle that is notched like a saw. Only the upper section of the branch is harvested, that year's new growth, red with resin and still flexible. You have to avoid cutting too low, at the woodier part of the stem, which is difficult to break and not productive. It is an impressive sight to watch experienced cutters in action. They grasp several twigs at once and use the sickle to break the branches as much as cut through them. They work quickly, very quickly. The handfuls of branches pile up on the ground until there are enough to make a bundle. Hanging from the cutter's belt are strings with which to bind them. Hunched over in the morning sun, the harvesters advance steadily through the field, using a pitchfork to load up their bundles, which spill over the edges of a cart harnessed to a donkey. It is a ritual harking back to the harvesting and haystacks of so many rural areas in France, a ritual that disappeared there fifty years ago. Yet here it is an enduring feature of farming life, no matter that cistus is tougher to slash than wheat.

The carts are unloaded next to the cans. The women have

made preparations for boiling the cistus, a process that will continue into the evening. Sticks that have been boiled up in previous days are used now as fuel to heat the receptacles, which are full of water and soda: the piles stacked up next to the cans are set alight. It is an astonishing spectacle in the afternoon heat, flames and smoke billowing up in the sunlight as the contents of the blackened drums are brought to a boil. The women use pitchforks to throw on the bundles that have been slashed that morning. After it has been boiling for an hour and the gum from the branches and leaves has become liquid, the fire can be put out and the boughs removed. Now for the most delicate part of the operation, a job for the head of the family. Wearing shorts, flip-flops and a shirt flecked with gum, he takes a drum of sulfuric acid and starts pouring it slowly into a bucket which he will then empty into each barrel. Everything smokes and bubbles as the acid neutralizes the contents of the container and precipitates the gum. A thick pancake of labdanum gum appears at the bottom of the barrel. It is stirred with a rod to remove the water and air, until it takes on the consistency of a beautiful beige butter.

Witnessing these enthralling scenes from another age, I could sense beneath this man's apparent nonchalance the silent legacy of generations for whom life had always been harsh, its risks a form of gambling with fate. By day's end, the two or three barrels that had been filled would be delivered to our production unit. And after drying out, the gum would be transformed into derivative products with their prized notes. The cistus has such a powerful scent that it would linger on the harvesters throughout the summer, and it would accompany me on my return to Landes.

---

Soon the story of the Roma gum boilers will be nothing more than a memory. The waste water, the fires lit at the height of summer, the use of acid and soda, the lack of any health and safety regulations, none of it can last forever. Municipal and regional authorities have been progressively regulating production and some local businesses now produce labdanum in safe workshops that recycle the water used in the process. There are still many Roma producing gum but one day they will have to be satisfied with merely harvesting the bundles of branches, work that is difficult but well paid. Recently, the Roma have been joined in their cistus domains by Romanians, who first arrived to pick the strawberries and oranges growing along the Huelva coast and were then tempted to head up into the hills in search of better pay. An affecting meeting of Roma communities whose shared heritage is buried too deep in the past at this point for them to have any sense of its existence.

Juan Lorenzo would often ask me how the gum or essence from his branches would find its way into the flacons of luxury perfumes. "Will you tell people about us in Paris and New York?" he would ask. "You'll have to bring the perfumers here to see us, I'd be only too happy to show them why Andévalo is the most beautiful place on earth." I readily assured him I would do so, but I couldn't admit to Juan Lorenzo that I knew the perfumers no better than he did . . . My company was in Landes, a long way from Grasse or Geneva, I knew not much at all about the industry, its machinations, its main actors. I would play the part, dropping a few brand names, and being French afforded me a degree of prestige, an illusion I did my best to foster and maintain. In time, and following the facility's success, perfumers did eventually come to Puebla and Juan Lorenzo was very grateful to me as a result. What a sight he was to behold: bright-eyed and sporting an impeccable cap, guiding our wide-eyed guests over to the gum extraction sites, where they could see the harvesters at work. In the evenings, the *jamón* from his farm wound up making him the star of the show.

Puebla de Guzmán is famous for its *romeria*, the annual pilgrimage which takes place at the end of April to celebrate its patron saint, the Virgen de la Peña. I had been hearing about it ever since my arrival. With its tens of thousands of pilgrims who come from all over Andalusia, and the hundreds of men and women on horseback, it was the pride and joy of the village, its *raison d'être.* A year after I met Juan Lorenzo, he invited me to participate formally in the pilgrimage, which involved my dressing up in traditional Andalusian costume and riding on horseback for two days. On the morning of the festivities, we all assembled on our horses to make our way in procession up the few kilometers of cistus-lined path to the peak crowned by the Virgin's chapel. The women were dressed in their riding habits if riding sidesaddle, in their Sevillian skirts if mounted behind their partners. For my part, it was as though I were in a costume drama, astride a fine-looking horse and wearing a flat-crowned sombrero, a grey waistcoat and leather chaps. I fell in behind Juan Lorenzo; our procession diverged from the road as it climbed, an elegant and colorful caravan making its way silently along the path through the eucalyptus and cistus. Arriving at the hermitage, the riders dismounted and hitched their horses in the shade of the evergreen oaks. The statue of the Virgin is brought out only once a year, and is carried by a dozen people who are selected through a complicated and rigorous process, an honor for which one must sometimes wait ten or so years. As the hours went by, the square before the chapel filled with thousands of people and when, finally, the bearers of the statue appeared, the atmosphere was at fever pitch. There were tears, prayers, people were singing, trying to touch it, the procession was barely able to advance. There was an air of unreality to the whole scene that I found overwhelming. Every aspect of local life and culture to which I had been exposed these last months now took on meaning against the backdrop of this spectacular ceremony, a timeless, otherworldly scene. It was as if the days I had spent among the cistus plants had all been leading me to this point.

We managed to draw closer. The Virgin, an imposing statue in sumptuous robes, was seated on her throne, held aloft at shoulder height. And there, on the dark crimson of her gold-trimmed cloak, a large cistus flower, its splendor on show for all to see. Large white petals at the end of a golden branch, and in the middle of each petal, five red spots: the "tears of Christ."

I was entranced. Here atop the hill, the flower on the cloak was an emblem, a physical manifestation of the bewitching scent of cistus fields in the summer, when the air is quiveringly alive and the leaves gleam in the sunshine with their layer of gum, a fine skin of imaginary molten metal from the overheated, mined earth of the Andévalo.

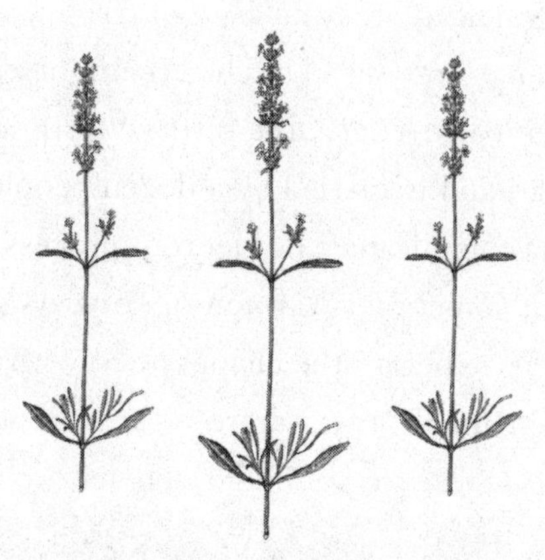

# BLUE HARVEST

*Lavender in Haute Provence*

"I have been smelling lavender ever since I was a kid, but this . . . ? I don't think I've ever smelled anything so good." In his Neuilly office of glass, aluminum and heavy carpet, Fabrice, the perfumer, takes his time to consider it. He is holding a tester strip, dipping the tip of the strip of paper into a small vial of essence. He passes it slowly under his nose, backward and forward, puts it down and picks it up, all in silence. Scent strips are the link between vial and nose, they are a perfumer's basic tool, a first impression before he or she smells a scent on the skin. I watch him concentrating on the new sample I have brought him. Fabrice is originally from Grasse and a renowned perfumer, a specialist in natural raw materials who divides his time between Paris and his beloved home town. He is inclined to reticence on account of his shyness, but his pale blue eyes light up at the jolt caused by a new scent. In Neuilly, he

*Jérôme in his lavender kingdom in Haute Provence*

is part of our company's fine fragrance creative team. In Grasse, he assesses the new notes being developed in our laboratories. Whether considering new plants or new extraction methods, Fabrice is the olfactory judge of all these ideas. His desk is covered with little glass vials, dozens of trials that are conducted every day for his many projects, all weighed and blended by robots.

Perfumers can be working, either alone or in a team, on numerous perfume compositions at any one time. They respond to "briefs," the description by a brand of the note it is looking for in the next perfume it plans to launch. Their formulae are complex compositions, subtle combinations of dozens of components, both natural and synthetic. Each ingredient is chemically and olfactorily mapped, each must conform to a precise aromatic imprint which perfumers keep stored in their memory, season after season. The balance of a formula's composition must not be disturbed by any possible variations in quality, and this often makes my task of procurement a delicate one. Quality and stability are two baseline requirements for a purchaser of natural products, come wind or rain. Perfumers must be able to modify their initial concept at a client's request over and over again before they are awarded the project. Frustration and disappointment are as much a part of their daily lot as the prestige of being a "nose" accorded to them by glossy magazines and the general public.

Fabrice is known for his creations for labels such as Diptyque, Réminiscence and L'Artisan Parfumeur, all brands with a reputation for top-of-the-range perfumery. They have benefited from his skillful structuring of natural materials, a talent which has also brought him great success at Paco Rabanne, Jean Paul Gaultier and Azzaro. He has taught me a great deal about how to smell. In the absence of any real apprenticeship or years of experience in the profession, it remains a somewhat hopeless quest on my part, but I have acquired the basics. In fields and workshops, I have learned how to smell the green or sweet facets in a floral scent, how to recognize the boiled notes in fresh essences, how to feel comfortable using vocabulary that is more evocative than descriptive. We talk of metallic notes, of humus, of rain, of cut hay and the cowshed, notes of salty skin, of new leather . . . Fabrice has given me a precious little reservoir of knowledge that I carry with me everywhere.

On this particular day, the two of us were discussing lavender, a flower so familiar but one that is constantly asking to be revisited. In the July heat of the Provençal sun, it has the intense scent of perhaps the most recognizable, the most accessible of fragrances, the perfume that evokes summer, linen cupboards, the freshness of eau de Cologne. Lavender is France's favorite scent, symbol of Provence, perfume of the Mediterranean south. Beneath the insolent certainty of a vast blue sky, the lavender fields are an ever-fluctuating palette of color, never truly blue, never entirely mauve. The change is subtle, shifting with the sun, the time of day, the orientation and size of the fields. Despite now being grown in many places throughout the world, its real roots are here, deep in this Provençal soil. Throughout the ages, lavender has remained the flagbearer of French aromatic products. Universally appreciated, it is a scent recognized by everyone.

When Fabrice talks about these spiky flowers, the sparkle in his eyes and his Provençal lilt remind me of the skies of Haute Provence. "Beautiful lavender is aromatic, fresh, sharp and vibrant. It smells clean, it is the smell of sun on white linen." We both know that these days Bulgaria has become the main supplier of lavender essential oil to the fragrance industry, hastening the decline in French production. It is a harsh reality for a child of Provence. "I regularly test the Bulgarian products, but most of them are fairly flat, with hints of fungus, Roquefort almost. This lavender you've given me to smell today is precise, noble. Where is it from, this sample of yours?" So I tell him. I tell him how the Provençal producers are desperate to save the French lavender industry, which is being threatened by cheaper, foreign competitors; how I met Jérôme, who has been growing a new hybrid for three years and who asked me, hopefully, if I might be able to put his samples in front of our creators. Fabrice adores it. "It's fabulous, I absolutely have to see this . . ." We make up our minds then and there. We will head south, to Manosque, where we will pay a visit to Jérôme and his fields. On his shelves, next to the flacons containing his best-known scents, Fabrice has displayed some old photographs of the jasmine, tuberose and rose harvests in Grasse. There is another of an old lavender alembic still sitting on a cart. In his mind, this son of a perfumer sees himself both as heir to and protagonist in this long history. To some extent, here in Paris, Fabrice is in exile.

For me, the grandson of a Provençal grandmother, a visit to Manosque takes me back to my childhood memories of holidays in the south, of a house whose every cupboard smelled of lavender. As a schoolgirl in Digne, my grandmother had lived through the glory days of this essential oil in the early years of the century. When she talked of those days, her accent from that time would reappear. She would tell us about the lavender, as well as the olive branches and candied fruit on Easter Sunday. Prior to the Great War, after the requisite moral instruction class, primary school teachers would ask their students to go home and tell their parents they must plant lavender. Families fell in behind a true regional cause. Of course, Manosque is also the setting for the work of its famous son, the author Jean Giono. In *Provence*, he writes that "fine lavender," the soul of Haute Provence, grows at higher altitudes, in the foothills of the Lure Mountain. He situates its historic heart in the poor sheep-grazing areas of the stony, windswept country between the Alps and Provence. In those early decades of the twentieth century, an entire region found itself dependent on lavender. A world of crops and alembics, of businesses in Digne and Manosque that would trade in the essential oil. In the words of Giono, "Evenings at harvest time are fragrant, the sunset colored by the litter of cut flowers, rudimentary stills set up next to the tanks breathe red flames into the night."

The history of lavender goes back further still. The families of shepherds in ancient times were already using sickles to harvest lavender from *baïassières*, great expanses of wild lavender that grew on the mountainside. Early alembics date back to the seventeenth century. And by 1850, the demand for lavender essential oil was so great that the distillation process had been transformed and expanded. The small, primitive stills along the edges of the fields were replaced by a mobile model that moved from village to village throughout the harvest season, distilling the sheafs of lavender as they were brought in by the farmers. For almost a century these stills formed part of the backdrop to everyday life in the region. Mules pulling carts bearing copper tanks gradually gave way to trucks employed for the same purpose. Lavender crops started to be planted in about 1890, a necessary development to meet industry's demand. Wild harvesting would not survive the labor shortage in the mountains after the horrific blood-letting of the Great War. My grandmother's brother, Julien, was killed at the Somme in 1915 at the age of twenty. It was a subject she never broached, preferring to share her lavender memories.

As crops were planted, lavender spears gradually disappeared from the mountains and their scent altered. The essential oil lost a little of its soul, and its smell was considered less "elegant." A ransom, of sorts, for its immense success, which for a century had seen lavender intimately linked to the growth of the Grasse fragrance industry, its prosperity inseparable from the spectacular success of the town's entrepreneurs and their perfume houses. The years 1920–30 marked its apogee, the golden age of Grasse itself and, at the same time, of its natural raw materials. The big names—Schimmel, Lautier and Chiris—would be associated with the town's name until the 1960s. In order to secure their supply of essential oil, all of them set up large distilleries in Haute Provence. Thanks in large part to its lavender, Grasse emerged as the perfume capital of the world.

---

After arriving in Manosque, Fabrice and I first head up to Valensole. Prior to the Second World War, this great plateau was nothing more than an expanse of stones punctuated by stands of oaks and pine trees, grazing sheep and almond orchards. When the almond trees on the plateau blossomed in February, it was a delight to behold, but frosts would ruin the harvest at a rate of almost one year in every three. Almonds remained a pauper's product; my grandmother remembers how the women employed to crack the nuts would be paid in husks, the by-product of their labor, which they would gather up to use as heating fuel. These almonds were initially intended for use by nougat manufacturers, and we would always spare a thought for these women and their work whenever we bought any nougat in Montélimar on visits to my grandparents in the south of France. Just before the war, some local entrepreneurs came up with the idea of converting the plateau into farmland and the story goes that, in 1938, the tractors brought into Valensole were among the first to be used in French agriculture. From 1950 came the recognition that these flat, desolate expanses could be used to establish significant lavender and wheat holdings. And within just a few years, the almond trees were pulled out, thousands of hectares were cleared of rocks, and Valensole was covered in vast fields of scented stalks.

Half a century later, all this has changed, although most visitors remain ignorant of the development. In a complete turnaround, the fields of Valensole—made synonymous with lavender in all-too-familiar postcards—have now been planted with lavandin. A cousin of "true" lavender, and a hybrid of two more productive and more resistant species, lavandin has colonized the plateau that it has dominated since the 1970s. Its oil is considerably cheaper than that of lavender and it has become known in the industry as a natural source of evocative notes notwithstanding its distinctly more camphorous tones. Lavandin is used in functional fragrances, for detergents, washing-up liquids and shampoos. The names "lavender" and "lavandin" are often used interchangeably, a deliberate conflation so as not to disappoint the tourists. The two plants look very similar; it takes a little practice to distinguish them. Lavender stalks are shorter and it tends to be bluer in color, retaining the distinction of its prestigious history and the elegance of its scent. Lavandin, however, has become an indispensable part of industrial fragrance manufacturing and these days prevails in the region. In order to track down real lavender, one has to gain some altitude and head up to where it was first grown.

In July of every year, between the valley of the Durance river and the gorges of Verdon, thousands of flowering hectares provide a unique spectacle, a spread of intense color, of undulating mauve and violet stretching to the horizon, where it meets the azure sky. Come mid-July, when it is time to bring in the lavandin crop, harvesters work day and night, tractors making an imposing sight as they enter the ocean of bluish-purple stalks, leaving behind them the pale-green stripe of shorn clumps. The harvested flowers are blown directly into containers that will serve as stills, hooked up to the steam generated by a nearby distillery.

Making our way slowly through this profusion of colors, Fabrice and I climb up past the plateau along little roads wending their way through the oaks until we reach Jérôme's farm, between

Mont Ventoux and Banon. Here is where you will still find Provençal lavender, confined to the high plains, far above the fields of lavandin. The son and grandson of lavender farmers, Jérôme is mid-harvest on this July morning but he gives us a hearty welcome. Bands of vivid bluish-purple perch atop beds of white-blond pebbles, and fields abuzz with bees offer a spectacular view over the valley toward the majestic Mont Ventoux. Reaching us on the gentle breeze is the sound of a combine harvester working below. Fabrice and I share a glance, and a moment of complete serenity.

"I know what this plant has meant to the people in these parts. Too often we forget," Jérôme says, and Fabrice, the boy from Grasse, nods his head. This young man has made a deliberate decision to continue its cultivation, and he is convinced the market will recognize the superiority of his oil over the Bulgarian product. Fabrice's arrival has validated his decision. He has gone through the process of receiving organic certification, aiming for quality and a top-end product, and has just invested in a new distillery. He is betting on diversification so has planted sage, thyme, helichrysum and, most importantly, a refined selection of lavender. For three years he has been one of the pioneer growers of the new variety which I put under Fabrice's nose in Paris, and it is a treasure which this farmer of essential oils is proud to show us. We will be buying his entire production; his investments and tenacity are paying off.

We take a walk along a ridge above the farmhouse; the field is secluded, a little hidden, and I take in the view to the distant Alps. This is his parcel of land, twenty lines running down the slope, a seductive, geometric pattern of blue standing out against the green hues of the distant valleys below. “It’s almost ready,” says Jérôme, rubbing the flowering stalks before smelling them.

“A pure, deep scent, no camphorous notes,” Fabrice says soberly. “Your lavender smells of the mountain breeze, that’s the difference.” The perfumer scans the rows, his unusually blue eyes focused on the Alps; he is immersed in the lavender, his nose setting off his creative impulses. “Lavender is not particularly fashionable in perfumery, but this note brings back the original subtlety of the oil.” Fabrice is thinking of using it for a project in its final stages. The producer can’t hide his pleasure at receiving a visit from a perfumer and imagining the fruits of his labor making their way into the flacon of a designer brand. It is now just a question of volume: will Jérôme’s crop yield sufficient quantities to allow it to be used in the formula of a new perfume? The vocabulary of the conversation between these two passionate Provençal men steeped in such similar culture is timeless, their collaborative spirit as harvester and perfumer evident. In Paris, a billboard campaign is advertising Fabrice’s most recent creation for Azzaro, but here he is, wandering the rows of lavender, determined to find fresh inspiration in the beauty of that field. Far removed from the bubble of his industry fame, he is continuing to write, with Jérôme, the age-old story of the fragrances of Haute Provence. Perhaps this is the real motivation behind my efforts, behind what I always struggle to call a profession: to be a “go-between” of sorts between growers and perfumers.

At the height of the season, the soundtrack of bees gathering pollen is intense. Fabrice is finding it increasingly difficult to mask his emotion in the face of the unusual beauty surrounding us. In these mountains, Jérôme explains to us, a farmer is also responsible for fashioning the countryside. With his lavender, his oaks and his beehives, he has a dream of winding the clock back a hundred years and perhaps preserving an entire cultural legacy. Colors and smells: a silent symphony is carried on the breeze. But Jérôme has no time for nostalgia on his grandfather's land. We spend a few hours discussing matters back at the farm; I make an offer to buy his entire crop based on his predictions for this year's harvest. A few months later, his lavender will occupy pride of place in one of Fabrice's beautiful creations for L'Artisan Parfumeur. It was Jérôme's lavender, Fabrice will confide to me, which resonated with him and spurred him to evoke the Provençal landscape in his fragrance Bucolique de Provence.

While lavender is making a comeback in the mountains, there is an abundance of lavandin on the plateau; the original plants from Haute Provence are these days destined for different fates. We walk back across the plateau in the late afternoon. Beyond the village of Valensole, several tourist buses are parked on the side of the road in the middle of nowhere, lost in a sea of lavandin. Twenty or so couples in wedding garb emerge from the buses in a surreal procession. Chinese women, dressed in white and holding parasols, laugh as they step out into the rows of violet flowers, mobile phones at the ready to capture the shot. Some years ago, two hundred million fans watched episodes of the Chinese television show *Dreams behind a Crystal Curtain*, in which the stars of the show were married in Provence. These days tourists from China come for an authentic experience in the lavandin fields, taking selfies, picking bouquets, smiling against a backdrop of deep purple stems. The symbol of modern agriculture in Haute Provence, lavandin now receives wave after wave of this modern form of tourism in an improbable blend of white and purple.

The next day, after returning from Jérôme's farm where we had walked with him through his stands of oak trees, I am reminded of the story of *The Man Who Planted Trees*. Set in 1913, this short story by Jean Giono starts with a sober description of the bleak Provençal high country where the only thing growing is wild lavender. A shepherd makes his way across the land, an iron rod for a staff, his pockets filled with acorns. Entirely on his own, he will sow the seeds of a forest on these moors, a forest that will grow, as Giono recounts, and transform the landscape. What will become of this part of the world now as it turns from lavender blue to lavandin mauve, as almond trees disappear but ever more tourists arrive?

To venture up onto these plateaux is to believe in Giono's words and in the picture of the landscape he paints, a place hidden away, beyond the reach of the tourist buses. With his oaks and lavender, Jérôme is heir to this story, a producer of essences that no longer grow wild, but whose perfume remains unique. And whose plants are to be found growing alongside trees that have set down roots where the writer, wandering that barren land, imagined them into existence.

# ROSE OF THE FOUR WINDS

*Persia, India, Turkey and Morocco*

I have been working with the scented rose for twenty years, a rose that is unique among so many thousands of ornamentals. I have been a grower, a distiller, a sourcing agent and purchaser of its essential oil. I have encountered it in all those countries where it blooms, having taken root along ancient silk roads and spice routes. Rose is the incarnation of perfume in the universal imagination; without rose there is no perfumery. It was revered by the Ancients in its every manifestation, be it fresh blooms or dried flowers, as scented oils, in fountains or in flavored wines. Over time, one particular rose became known as *the* rose of perfumes: *Rosa damascena*, originally from the region around Shiraz, in Iran. Having made its way from Persia along the known world's trade routes, the rose arrived in Damascus, that great Mediterranean trading hub of the Middle Ages, and from there it was brought to

Europe by the Crusaders and christened the "Damask rose." With their invention of rose water in the eighth century, the Persians went on to perfume the world from China to Europe for eight or nine centuries until the discovery of rose essence, in India, in the seventeenth century, led to the transformation of rose into a perfume.

My mind floats back through the roses I have seen, carried on the four winds. Brief encounters or longer stays, how I have delighted in inhaling their scent wherever they may have taken root, their seeds carried away by the caravans of history, far from Shiraz. The rose is always seductive, no matter where I have encountered her, a princess in secluded gardens, in isolated mountain landscapes or growing at the desert's edge. Wherever she grows, there is running water to be found; look at her, surrounded by poplars, walnuts, and other fruit trees, swaying in the wind alongside fields of wheat or alfalfa, beneath swooping swallows and trilling nightingales. The young women who pick the roses can't resist using them to decorate their hair. For the gardeners who tend them it is a chance to inhale their scent every morning and smell the fragrance of the rivulets of essence that will collect in their stills. Every spring the rose bursts into a delicate pink frenzy before once again retreating from the world.

Persia has always had a love affair with roses. For more than a thousand years, the flower has been an integral part of the country's history and culture, occupying a fond place in the hearts of its people. I first went to pay my respects to its birthplace in Shiraz, city of the rose and the nightingale, two symbols so enduringly intertwined in Persian poetry. Later, in the bazaar in Isfahan,

*Rose distillation in Hatras, India: not much has changed since the time of the Mughals*

surrounded by every spice in the world, I came across dried rosebuds, a deep, almost-violet pink, smelling of both blossom and hay. The vendors were also selling a range of traditional bottles or flacons of rose water with labels that competed in brightness. In Qamsar, the manufacturing capital of rose water in Iran, I saw dozens of small producers who had set up in the courtyard of their home, distilling flowers in rudimentary copper stills. The recipe for rose water is as ancient as it is simple: a mixture of fresh flowers and water is boiled and the collected vapor then condensed by passing it through cold water. The water-soluble essence from the flowers is captured in the steam and perfumes the collected water. Sometimes a fine skin of golden essential oil can be seen floating at the neck of convex glass bottles of rose water, the insoluble essence being a mark of quality. Rose water is ubiquitous in Islamic cultures; valued for its purifying properties, it is used for sprinkling the walls of houses and mosques as much as it is for hand-washing. In Iran it is a part of everyday life.

I have traversed the Iranian plateau, a windswept ocean of minerals bordered by distant blue mountain ranges, a procession of pistachio crops, pomegranate orchards and mudbrick villages hunkered in the shade of jujube trees. From north to south, I have been enthralled by the fields of roses, ribbons of green in the desert, adorned with flowers made particularly vivid by the altitude and dry air. Planted high in the mountains at over two thousand meters, rose bushes laden with buds swayed in the winds in absolute silence.

Once I met with some rose farmers at the end of a track in the desert at what looked like an oasis in the middle of nowhere. As I sat drinking tea around a fire in the evening, I saw that with the exception of the little radio next to the teapot, little had changed since the time of the caravans. A bird started to sing in a jujube tree above the crackling fire. It was a bulbul, the nightingale, of course. For more than a thousand years, nightingales have been singing in the rose gardens of Persia, as scented water runs gently through her veins.

The story surrounding the discovery of rose essential oil—the substance that has been finding its way into our perfumes for four centuries—is a fascinating tale. In 1611, in the northern Indian city of Agra, the Mughal emperor Jahangir celebrated his wedding to Nur Jahan, a Persian woman of uncommon beauty and intelligence. Princess Nur, alerted by her mother, is said to have noticed the formation of a golden film of oil on the surface of the hot rose water baths prepared for the festivities, thus leading to the discovery of rose essence. She presented her husband with the precious liquid and he wrote: "Here is a fragrance so potent that a single drop on a palm can perfume the entire room as if a ton of rosebuds had all blossomed at once. It is unmatched by any other scent; it comforts our hearts and nourishes our souls."

Three hours' drive from Agra and the Taj Mahal, I went searching for the traces of Nur's essence in a distillery where nothing appeared to have changed since the time of the Mughals, except perhaps for a few electric light bulbs. In a large farmhouse built of rammed earth where everybody worked barefoot, in loincloths and turbans, a distiller was hunched over a great copper vat, manually molding the clay connection which would link up to his still. Bamboo canes were roped together in an elaborate braided design, a veritable work of art. The oil itself was collected in finely worked copper vessels and stored in a cool, clay-walled room. Beneath the stills the fires were fed with dried cow dung. There was something majestic, almost mystical, about these distilleries that dated back to the building of the Taj Mahal itself, their flames a silent homage to Jahangir and Princess Nur and their discovery of the essence of rose.

In Turkey I spent several years setting up a factory to produce rose extracts. Since the 1930s, fifty-odd villages around the city of Isparta have been responsible for growing the country's perfumed roses. The Turks spent almost fifty years trying to bring back the roses lost by the Ottomans when Bulgaria, home to the sultans' preferred rose gardens, gained its independence. I remember Ahmed, our flower broker in his remote village, deep in a valley. The surrounding hills were covered with plots where meticulously maintained rose bushes clung to the slopes like carpets spread out between wheat fields and apricot trees. Nestling in the shade afforded by walnut trees were farmhouses made of stone, cob and wood. The women worked at their looms and in the fields, men chatted in the café, smoking endless cigarettes, drinking tea and playing dominoes. Ahmed's shop was a small, limewashed room painted blue, equipped with a table and a set of scales. On the wall hung a sepia portrait of Mustafa Kemal Atatürk, the nation's father, with his lambskin fez and wolf-grey eyes. It was he who was responsible for kick-starting the rose industry in Isparta in the 1920s, which led to the establishment of a large cooperative and several distilleries. Ahmed invited me to lunch on his wooden deck, where we sat in the shade of an ancient walnut tree and he introduced me to his youngest daughter. Songül must have been ten years old, a little girl with an intense gaze whose name, I was told, meant "the last rose." She seemed the perfect manifestation of the Turks' determination to continue their cultivation of the gardens once so beloved by the sultans, and an incarnation of the Ottoman pride that their land was being used to distill this queen of flowers.

Far from Shiraz, in southern Morocco, at the foot of the Atlas Mountains, Damask rose bushes bloom every April. Nobody knows any longer when or why these roses arrived and prospered here. At the end of the 1930s, the French colonial settlers established two flower extraction plants in the small town of El Kelaâ. They had noticed that the farmers would plant rose hedges around their crops and then harvest the buds, dry them and use them in their henna. These desert factories were, and still are, extraordinary. Built on stone and sand, they sit within a fortified structure, the *ksar*, a vast courtyard surrounded by crenelated walls and corner towers. They look out over the green crops growing along the river below, and have a striking view to the Atlas Mountains beyond. For years it was my job to monitor the quality of the crops grown on the land which supplied our facility, visits which had an almost dream-like quality as I was plunged back in time. The extractors were located in the workshop. Great, black, cast-iron wheels that looked much like giant washing machines. Everything remained as it had been when the factory was first built fifty years earlier, the big fuel-oil boiler along with the enormous strongbox. The atmosphere of the place harked back to the 1950s with its ancient, handwritten ledgers recording flower purchases and production, flacons bearing the names of companies that no longer existed, and a fit-out to match.

Leaving the factories to head down to the hedges of roses, one comes to the verdant banks of the two rivers that run through the valley, a lush mosaic in the middle of the desert. Depending on the season, water is channeled through tiny canals into the small fields of broad beans surrounded by rose bushes and fruit trees. In the early mornings, young women in traditional Berber dress, faces hidden by shawls and hats to protect them from the sun, work their way along the hedgerows with their baskets. Quickly and efficiently, they collect the roses, which could almost be described as growing wild. Scattered through the fields are the imposing silhouettes of the *ksour*, built at the water's edge so long ago. They have all been abandoned, these desert castles, their red-ocher earth left to glow splendidly in the sun. As their roofs collapse, the *ksour* are starting to disintegrate in the rain; there is no sight more melancholy than these ruins, holding out against their garden of Eden backdrop, their rammed-earth walls slowly, almost regretfully, dissolving. The silence is disturbed only by the sound of birds and water. Wind rustles through the willows, children wander past, nudging a few cows ahead of them, followed by old women balancing enormous bales of alfalfa on their heads, while the younger women carry the harvested flowers back to a weighing station.

I like to think that the Damask rose stopped here to set down roots, having found an ultimate oasis of beauty. That is, before discovering Bulgaria, whose national emblem it is.

# THE BIRDS OF SHIPKA

*Bulgarian rose*

My first encounter with the Bulgarian rose dates back to 1994, less than five years after the fall of the Wall and the official demise of the communist regime. I was attending an international conference organized by the state monopoly, Bulgarska Roza, the sole national producer and vendor of rose essential oil. In Kazanlak, a town located in the center of the country—historical capital of this queen of flowers—the first thing suggested to the occasional foreign visitor would be a visit to the Rose Museum.

Situated a little outside the town, the state-run Rose Institute now found itself short on resources in these changing times, but was still managing somehow to employ a small team of agronomists and to maintain some crops of aromatic plants and, most importantly, its museum. What an extraordinary experience! A suspicious woman—the building's caretaker—reluctantly opened the deserted space and showed us down to the cellars where some exhibits in a few damp rooms struggled to retrace four centuries of the flower's noble history in Bulgaria. Visits were infrequent, that much was clear, and it was moving to see the general state of obsolescence: collections of old photographs of the first distilleries dating back to 1860 with their range of little wood-fired stills, the first large exporters posing proudly in their laboratories with medals won at the perfume trade fairs of Vienna, Paris and London. Handwritten ledgers recorded the production of rose oil from each village in the valley at the start of the century. I came across *konkums*, which had been used for export for as long as anybody could remember, a sort of round, flattish flask, initially made of copper and then of tin, which would be delivered to purchasers wrapped in a cloth and ribbon in the Bulgarian national colors, and topped with a wax seal. The museum had an unusual 200-liter *konkum* on display, a unique object that still smelled of rose fifty years after it had been emptied. As the visit progressed, the story of a golden age emerged, a perplexing contrast to the current fusty presentation. It was hard to believe that our Bulgarian guide felt any real pride in her country's early manifestation of capitalism in the early twentieth century. The first heroes of the Bulgarian rose oil industry were the entrepreneurs of the late 1800s, whose fortunes suffered a fatal blow when the communists came to power in 1947. Our guide had much more to say in the second half of our visit, which focused on the post-war glory years, vast fields and large tractors, teams of flower pickers, modernization and state-run factories. The icing on the cake was the exhibit of photographs from the Rose Festivals in the 1970s,

*A rose picker in Bulgaria, five thousand flowers in her bag will give four or five grams of oil*

especially the gallery with the Rose Queen portraits. And that is where the display finished, as if time itself had come to a standstill. The guide's answers to my questions about the current state of rose oil production were as perfunctory as they were vague. There were three state-run companies; they produced the best essential oil in the world because the Bulgarians continued to be the leading experts in both the cultivation of the flowers and their distillation. I did not have the heart to ask her why the Bulgarian rose had disappeared from international markets and from perfumers' formulae, having been replaced by Turkish oils.

The museum was selling a booklet that set out the origin and history of perfumed roses in Bulgaria and went some way to explaining why this centuries-old tradition was such an integral part of the country's cultural heritage. Plantations in the Rose Valley date back to the seventeenth century. The Ottoman Empire's demand for rose water and essence continued to grow, along with its reluctance to rely solely on supply from Persia, the birthplace of the *Rosa damascena* and the source of rose water since the year 1000. In the mid-sixteenth century, Sultan Murad III commissioned his gardener to develop rose plantations in Kazanlak, a town in his home province of Edirne, for the purpose of supplying the palace in Constantinople. This signaled a new era of wealth for the town, which became the source of rose products for the entire empire for almost three centuries. When the Bulgarians achieved their independence in 1880, they were keen to claim the invention of modern rose oil, having developed the technique of double distillation, which led to the production of the essence that is known and used by today's perfumers. For some sixty years through to the 1920s, the Bulgarian rose enjoyed an international reputation, a golden age which the museum was now struggling to piece together from people's memories.

The Kazanlak museum told the tale of two different stories: the very real account of a glorious century in which, for the world of perfumery, there was only one rose, namely the rose grown in the Kazanlak valley; and the touching efforts to paper over the current state of their industry's humiliating decline, which I was soon to witness first-hand.

It was shocking to see the state of disrepair in which Kazanlak found itself. The large armaments factory, Arsenal, which had provided employment for the town, had not recovered after the fall of the Soviet Union and hundreds of workers were now on the street. The only things the town had to offer were a few handsome but abandoned nineteenth-century Turkish houses and the linden trees lining its streets, a contrast to the rows of grey buildings, the wastelands of decrepit factories scattered through the town and the Stalinist concrete of the conference center. Only the sweet honeyed scent of linden blossoms in June could still evoke the town's heyday when, sixty years earlier, it had enjoyed its reputation as the rose oil capital of the world.

I have forgotten nothing of those first days I spent in Bulgaria. The conference was surreal; everything was directed at convincing the few foreigners present that the country's essential oil industry was flourishing. The stage had been carefully set: visits to decommissioned plants where a boiler lit that very morning gave the impression of producing steam and where a team of women employed for the day pretended to distill a few bags of flowers that had been hurriedly brought in. At the end of the day, the managers at Bulgarska Roza would give appropriate speeches about the international community, liberally accompanied by toasts to mutual friendship. One evening, I found myself enthralled by the enthusiastic post-prandial festivities that took place in a bear-hunting lodge, surrounded by beech trees and looking out over the valley, which belonged to the former leader of Bulgaria, Todor Zhivkov. Bulgarians are Slavs of the south, Mediterraneans who love to drink and dance. As the evening progressed, the formulaic speeches gave way to increasingly solemn renditions of traditional songs, until I had the feeling that an entire people was singing its history. The tears springing from the guests' eyes were not entirely attributable to the rakia, Bulgaria's national drink; they spoke also to the wounded pride of these people of the rose and their nostalgia for an era which they themselves had not experienced but which they all carried deep in their hearts.

The apparatchiks from Bulgarska Roza, having assessed who among their guests might be potential purchasers, did not leave my side. By the mid-1990s, rose oil production had all but ceased. The small quantities of essential oil being distilled would be sent back to the cellars of Sofia's Central Laboratory, the national treasury where samples of rose oil from every era were stored. Every technical and financial detail of the system was shrouded in secrecy. It would take years for the laboratory to resume normal operations after being the subject of endless rumors and fanciful theories surrounding quantity, quality and sales, which were, to a greater or lesser degree, state-controlled.

I was keen to see the plants and fields, a request that was not met with enthusiasm, but one which, in the end, was agreed to—provided, of course, a guide accompanied me. And so I was allocated Vessela, a young, French-speaking engineer who was entirely clear-eyed about her country. Her father had received a permit to work abroad, so she had lived for a number of years in Morocco as a child. She was now working for a pittance in an essential oils laboratory. She knew France and was fervently hoping to reinvigorate the fortunes of the Bulgarian rose, well aware, however, that it would only happen if the current systems were to change. While she was at university, the police had approached her discreetly to offer her a place on an elite training course. Realizing this would have been a stepping stone for a career as a spy, she had been brave enough to decline, knowing that, henceforth, the authorities would be keeping an eye on her. She remained resolutely interested in an opportunity in the West and had been waiting for her chance; our meeting had come at the right time. During the conference, and contrary to what had been expected of her, she had opted to provide me with a truthful account of the state of the country's rose industry. Thanks to her, I came to understand that the rose oil would only start to flow once again for whoever was willing to make the investment. I returned from Bulgaria convinced of the need to become a producer there and to invest in the industry—and, if at all possible, to be the first to do so. However, the arrival of a foreign company in the Valley of Roses was precisely what the Bulgarians themselves did *not* want to see.

I employed Vessela. She made sure I was introduced to Nikolaï, an agronomist who specialized in growing roses and who evidently was out of a job. Taciturn and very agreeable, Nikolaï would start any meal with a little glass of rakia that would loosen his tongue and get him talking about roses, a topic about which he knew all there was to know: where and how to plant them, their preferred soil, the correct exposure, the orientation of the rows of rose bushes vis-à-vis wind direction. He had particular experience with organizing several hundred pickers to harvest the flowers. The two of them, Vessela and Nikolaï, made the ideal team. On the one hand, there was Nikolaï, the invaluable technical expert, grumpy and anxious, and, on the other, Vessela, the intrepid optimist who knew how to carry everyone along with her enthusiasm and how to stare down adversity. As teenagers, they had worked together on the collective harvests as "volunteer" pickers along with every other high-school student in the country, harvesting tomatoes, sweet peppers and roses. Now they couldn't bear seeing the steady abandonment of the fields of roses. Together we planted our crops, built a distillery and started producing rose oil. We were accomplices in our shared romantic vision of saving the Bulgarian rose.

Our first rose harvest campaigns were the stuff of legend. In 1995, there was no question of foreigners purchasing a facility, nor even of setting up a company. The solution? To rent one of the existing state-run distilleries that had closed down when the money ran out. Nikolaï was our front man, which allowed us to take on the risk of one distilling season, provided we could find the flowers and put together a local technical team for the three-week period. As expected, we encountered many roadblocks. That first year, I was denied access by the police to the distillery we had rented because I was a foreigner, on the pretext that the Bulgarian technology was exclusive and confidential. In order to protect these trade secrets, two policemen took it in turns to guard the entrance, and even though there was no unpleasantness, I spent the picking season outside the factory. The distillery we had found had not been operational for the last five years. We had had to chase away chickens and recondition the old engine that would serve as a boiler. Copper stills, however, never grow old, and they still smelled a little of rose. Nikolaï had managed to purchase flowers to be harvested and had assembled teams of pickers, while Vessela persuaded a few veteran distillers to come and work for us. They were mostly unemployed women with no other means of support, nostalgic for the industry's glory days. They were all keen to work hard and we managed to produce twenty kilograms of rose oil, a result beyond our wildest dreams. Everything was in Nikolaï's name: lease, employment contracts, it was all Bulgarian. Exporting the essence to France was very challenging, but Vessela was able to work miracles. We were successful, and our initiative came as a bombshell within the little world of the rose. We had just carved ourselves out a corner of the monopoly, and everybody realized that nothing would ever be the same again. Vessela received threats and was declared a traitor to her country by officials in the industry. The obstruction continued for two or three years, until the first signs appeared of the economy opening up, and of opportunities and positions for Bulgarians in the new companies that were emerging.

The country changed a great deal over the next five years; private partnerships were formed to snap up everything that was being sold off by the state. Russian mafioso practices had taken hold throughout Bulgaria, but the rose industry was too limited to attract their attention or their appetite. One day in the year 2000, we visited a distillery in a small village largely inhabited by Roma families, east of Kazanlak, on the banks of the Tundzha River, which runs through the Valley of Roses. The countryside was magnificent, as much of the Bulgarian countryside can be: a paradise of fields and forests, of birds and wild eglantine roses. The factory was of course in a state of complete abandonment, but it came with ten large stills and a house nestled in the shade of cherry trees, lindens and walnut trees. Hundreds of swallows had taken up residence. We purchased it and renovated it from top to bottom; the swallows continued to make themselves happily at home. Every year they were a feature of the rose harvest campaigns in May and June.

In order to operate a distillery, you need flowers, lots of flowers. At least three tons of roses for one kilogram of rose oil, in other words one million roses, each flower hand-picked. To rely on buying in the flowers was a complicated and risky strategy, so we planted a hundred or so hectares of rose bushes. That winter, we had to mobilize two or three hundred villagers who were happy to have a little work. It was cold, the men all carried a bottle of rakia on them and there were many older women doing the hard work with a hoe to cover the seedlings. The younger workers piled up cartloads of rocks. Never had I been so aware of this act of plowing the earth being the primal act of a perfume's creation. We were cold, the women's hands were ruddied by the Balkan wind. I thought back to the crushing heat of my days in the cistus fields and of the men and women so arduously paving the way for the advent of such incomparable scents. The seedlings, which were barely poking through the soil, would be in full leaf some four to five months later, and would flower the following year. Many of these workers would return for the harvest, some working to fill the stills of our distillery. A long way from the cities, these Bulgarian villagers found themselves living hand to mouth in those years when state subsidies had disappeared along with the jobs at the Arsenal factory. The rose harvest, the cherry picking, every one of these seasonal jobs was eagerly awaited.

Our fields expanded, plots of fifteen or twenty hectares covering the gentle slopes of the valley. By the second year they were producing their first flowers, by the third our bushes were head height, and in no time at all we were able to start picking. We had become part of the Valley of Roses' long history. The valley itself covers a hundred or so kilometers; it is an ideal setting, given the light, well-drained soil, its moderate altitude and, in particular, its climate. Spring nights are cool, so mornings are damp and dewy, ensuring the buds do not open too quickly in the sunshine.

Picking the flowers is normally a job for those who live in the villages close to the fields. Many of them have a substantial Roma population. There are approximately one million Roma in Bulgaria, more than ten percent of the population, a sensitive and complex topic. The Roma live on the margins of society, and whether this is a situation imposed on them or a deliberate choice on their part is the subject of endless debate, in Bulgaria and elsewhere. The majority of the country's population considers itself to be Slavic, descendants of the Thracians, and they dislike the Romany people, whom they do not consider "Bulgarians."

The numerous Roma communities in the valley live mainly from picking and harvesting: mushrooms, chamomile, seasonal fruits and, of course, roses. Twenty-five years ago, the task of picking was primarily undertaken by local villagers, and in particular by the women who were reputed to be better pickers. Over the years, the villages have emptied out and now the harvesting is carried out almost exclusively by the Roma. They start early, working from six in the morning through to midday. A good picker will harvest forty-five kilograms in a morning: three bags, each containing five thousand flowers, plucked one by one between thumb and index finger. Women and men of all ages chatter and sing as they pick. One woman launches into a moving song. She tells me she is a Russian immigrant, and her song of the Volga reminds her of home. She works the fields in winter also, but it is too cold then for any singing. Now she sings for the roses. The Roma pick in small groups. The young ones are cheerful and joke around, the girls making crowns of roses for their hair. At the end of each row, ponies decorated with a red-pompom good-luck charm wait for their carts to be loaded with clear plastic bags filled with flowers gently warming up in the sun. The faster they are brought back to the processing plant, the better the yield of rose oil.

Nikolaï is everywhere, overseeing work in the fields. He has to manage hundreds of pickers, organize the teams, allocate the rows. Some days the Roma turn up. Some days they don't. If it rains, everybody is reluctant: the work is grueling but damp flowers weigh considerably more and payment is made by the kilo.

On days when the roses are in full flush, the sight of thousands of pink buds unfurling in the sun is a unique spectacle, and management can be a challenge. After seven o'clock in the morning, the buds "explode" and the fields become a sea of flowers that must, as a matter of urgency, be harvested before nightfall. By the next day, any roses that haven't been picked will have faded, the yellow stamens will have blackened and a large part of the fresh flowers' essential oil will have evaporated in the sun.

Pay day comes around every three days and each field has its weighing station, preferably situated under a walnut tree. A great deal of money is kept in the car parked under the tree, and the discreetly armed guards who have accompanied Nikolaï to the bank stand a little to the side, keeping their distance. The tension is evident in the general silence, everybody is waiting their turn, weighing tickets in hand, talking softly. Nikolaï does not take his eyes off the hands of the female supervisors rapidly counting out the wads of notes.

Every year, around May 20, the factory swings into its production campaign and Vessela reigns over her domain in the distillery. She will have recruited a team that works day and night for three weeks, even sleeping on site. Each season's campaign presents a fresh challenge. There must be a constant, uninterrupted supply of flowers, bags prepared and at the ready so no time is wasted filling the stills. The atmosphere is feverish for these weeks, the factory a real hive of activity, and Vessela is the queen. The work is accompanied by a hectic flurry of swallows, whose cheerful twittering is even louder on days when the yield is good. A distiller is responsible for one production line, that is, four stills linked to a single column. The column is the heart of the production process and is managed by experienced women who are proud of the expertise they acquired in the state-run factories before they were shut down. The trucks arriving from the fields are unloaded and the bags stacked up around each still, thirty-five of them ready to be tipped in for the next batch. On hand are young Roma boys, who pour the flowers into the mouths of large copper pots. The stills are steaming away, lending the factory a powerful smell of distilled roses, a blend of floral and spicy notes, almost peppery. It is a warm, fresh smell; the rose oil must now rest for several weeks in order to lose its "boiled" feel and reveal its true fragrance.

The work of distilling continues all day and carries on into the night if flowers remain from the day's harvest. In the middle of the picking campaign, there can be so many flowers that either batches must be increased or the distillation process abbreviated—difficult decisions that will have an impact on yield and quality.

Every morning sees the rose oil collection ceremony. Pipes from the distillation lines lead to a big "florentine" flask, a traditional vessel used in perfumery that allows the essential oil floating on the water's surface to be recovered. The florentine, which is the final receptacle of the prized essence, is hidden in a separate room where we shut ourselves away for the duration of the process with Vessela, Nikolaï and Nelly, the distillery's manager. The operation involves capturing the rose oil that has been distilled in the previous twenty-four hours. Nelly has placed a large jar under the florentine's tap. Wax seals ensure that the outlet tap through which the rose oil flows is not sullied, a protective device straight out of the nineteenth century. After a few minutes, a golden liquid appears at the top of the flask and rises up through a glass tube. The tension is palpable. Everything is critical from this point: the color of the oil, a pale yellow with glints of green; its clarity; and, of course, its quantity. The oil has started to flow, its scent pervading the room, heady and strong. The liquid runs slowly into the heavy glass receptacle that Nelly is holding in her arms. Nobody wants to interfere with this process; a little slice of the history of the rose in Bulgaria is being played out here every day, and it is a meaningful moment for all involved, as powerful as the scent enveloping us. These same gestures, this same ritual, the same silence all form part of a process that has endured unchanged throughout the ages. We have just extracted four liters of essence, everybody is smiling: today's yield is good. More importantly, we have just witnessed true alchemy, a process that started in the winter fields as the soil was transformed into flowers, continued with the flowers' harvest and distillation, and finished with their mysterious transformation into this golden liquid. The essence that Nelly is holding in her arms is worth as much as a gold ingot. An ingot weighing four million hand-picked roses.

After being weighed and filtrated, the new batch will join the others in a small, reinforced room. Plans for export following the picking season are kept confidential. The ten-kilogram lacquer-coated aluminum *estagnon* tanks take off from Sofia's airport on dates that Vessela has fixed and which have not been shared with anybody else at the factory. A small van arrives at dawn to load them up before heading to the airport under the constant surveillance of a pair of armed security guards. In the first years, the risk was so great that we had to resort to diversionary tactics. One vehicle loaded up with empty containers would set off, then, two hours later, a second vehicle carrying the precious merchandise would depart.

The picking campaign finishes mid-June, just as the lavender fields a little further to the east are starting to turn blue. The whole team celebrates the end of the harvest on the terrace of the house, eating Bulgarian cheese, cherries straight from the tree and strawberries from the next village. The rakia gets passed around. Profoundly exhausted, yet proud, Nikolaï takes a drag on his cigarette. He is spent. The Roma who have just finished the harvest make their way past the factory on their carts, waving broadly; they are heading off to the river to fish. This is our tenth season. Vessela is reminiscing about the early days and our encounters with the police in Kazanlak, and everything that has changed in the meantime. Now, in the mid-2000s, Bulgarian roses have started to bloom once more. There are fresh plantations throughout the valley, as well as new and restored distilleries. A great deal of money has been invested thanks to European Union subsidies, and the new producers reflect the new face of the country itself: mafiosi looking for ways to launder their money, property developers with an eye for easy returns, former stakeholders of the state-run companies that have since been privatized. But also young, ambitious entrepreneurs and a few passionate Bulgarians.

In the meantime, I have switched from being a producer of essences to being a purchaser. Every trip to Bulgaria involves a visit to our fields, where I have a bite to eat with Vessela and Nikolaï. I also drop in to see Filip, a former competitor who is now one of my suppliers; he is a dedicated producer and heir to a legacy that could, on its own, write the story of the Bulgarian rose. Filip's facility, the Enio Bonchev distillery, was established in 1909 in a small village not far from Kazanlak and was, at the time, the largest in the country. In order to respond to increasing demand from perfumers in Grasse, production in the Valley of Roses was structured around big companies equipped with steam boilers and large-capacity stills. Enio Bonchev was one such pioneer and the business was very successful until its nationalization by the new regime in 1947. Very quickly abandoned, the facility was saved by its idyllic surroundings, which led to it being turned into a museum. When I met Filip and his father, Dimitre, at a conference in 1994, they had just been successful in long-drawn-out restitution proceedings. As the sole representatives of a yet-to-be-established private production industry, they were regarded with considerable suspicion by directors of the state-owned companies.

Having been competitors for years, we now find ourselves partners. A passionate devotee of the rose, Filip now manages the family business, which has become a leader in the field. He has retained the historic part of the facility, with its copper alembics. The building nestles under great trees, some of which are as ancient as the building itself. His little museum displays beautiful photographs of the industry's glory days, and he has also preserved the cool rooms where the roses would have been spread out before distillation on those days when there had been a bumper harvest. Filip considers it his duty both to educate and to be a repository for the collective memories. He sells two or three grams of genuine essence in pretty wooden vials to tourists passing through, while railing against the synthetic products that are sold in Sofia as real rose oil. A leader of the new generation of producers of natural products throughout the world, he has no words harsh enough for those who cheat and dilute the essence. But that is a scourge which has been around a long time given that rose oil is literally worth its weight in gold. Even prior to 1900, cheap geranium oil was being mixed with rose essence, and newspapers of the day would run headlines about the scandals of what is known in the industry as the "adulteration" of essential oils. Fraudulent products abound and developments in chemistry make detecting them all the more difficult. But there is still one precious weapon in the arsenal: the trust between producer and purchaser.

Time passes slowly in the Valley of Roses, a landscape that has barely changed in the last hundred years. There are several accounts dating back to the nineteenth century of the emotion and wonder experienced by travelers from western Europe, when, after crossing Shipka Pass, they descended into the valley and came upon the silvery ribbon of the Tundzha River, then the dark green of the walnut trees, and finally the procession of pickers. Vessela and Nikolaï have continued to plant roses, and we like to reminisce when I visit. We remember having come across an interesting parcel of land fifteen years earlier in Shipka, the famous village not far from Kazanlak. Shipka was the site of the last battles to liberate Bulgaria that took place in 1878 between the Russians, who were supporting the Bulgarian partisans, and the Turks, who had been occupying that territory for the last five hundred years. An imposing Orthodox church was built in 1902 to honor the memory of the soldiers who had fallen. It looks out over the plain from its glorious site in the forest, golden onion domes peering out through the trees. Nikolaï started preparing the new plot for planting and we were very pleased with the outlook of these hectares of rose bushes below the church. One winter morning, we were walking together along the rows of newly planted young cuttings when he turned to me with that gentle gravity which is so typically Bulgarian. He told me he had a special present for me and from his pocket he pulled four uniform buttons. He had found them when clearing the site after the tractors had been through; they had belonged to Russian soldiers and had been lying there for more than a hundred and twenty years.

Sometime later, mid-harvest at the beginning of June, Nikolaï and I paid an early-morning visit to those flower beds. The field was a sublime sight: a broad slope running down to the valley that was covered in a sea of pink buds about to burst into bloom. As the sun caught the rose bushes, the birdsong, sparse at first, then louder and louder, took over the whole field. It was as if with their song they were encouraging the flowers, just beginning to open, still moist with dew, waiting to be picked. It was an astonishing sound, and I was surprised not to see a single bird. After a moment's silence, Nikolaï came over to me and said softly, "It's not birds you can hear. It's the souls of the soldiers who fell here, singing lest we forget."

# CALABRIAN BEAUTY

*The bergamot of Reggio*

While not a well-known fruit, bergamot has been delighting perfumers for three hundred years, thanks to the essential oil in its zest. It grows in the heart of the Mediterranean, along the Calabrian coast which looks across to Sicily, a land charged with a history so long that it is embedded in the 3,000-year-old myths of Homer.

I first encountered Calabrian bergamot when visiting the Strait of Messina, more than twenty years ago. My thirteen-year-old son had just finished reading the *Odyssey*, and he reminded me of the dreadful trials of Odysseus as he was forced to grapple with Scylla and Charybdis, the supposedly impassable guardians of the strait, monsters invented in part to illustrate the dangers of navigating a passage through those waters. My son had also just won an archery championship and was fascinated by the character of Telemachus,

*In Reggio Calabria, Gianfranco shows a* calabrese, *used by his grandfather to produce bergamot oil*

*A sea of bergamots on their way to be processed*

the son of Odysseus, and an accomplished archer. My recent travels through Bulgaria, Morocco and Madagascar were fueling his imagination. I was bound for one of the legendary lands in the *Odyssey*, and this, in turn, had led to wisecracks about similarities between this adventurer father of his and Homer's heroic protagonist. The truth of the matter was that this Odysseus-like father was off to negotiate the purchase of bergamot and lemon essential oils, and he would be back in a week.

Calabria is awash with history and stories. Tales of families, first and foremost, which themselves reflect the story of bergamot. On this February morning in 2018, I'm walking with a man by the name of Gianfranco along the esplanade in the provincial capital, Reggio Calabria. He is the region's largest producer of citrus essential oils. I had first come to see him twenty years earlier, when I had promised my son, otherwise known as Telemachus, that it would not be long before I returned home. I had been welcomed by Gianfranco, while his father continued to keep a watchful and expert eye on the running of the facility. Nowadays, people from these parts call him *Dottore* Gianfranco, in deference to his engineering qualifications and his successful career. A Calabrian from Reggio, an Italian and a European, Gianfranco is charming, elegant, a gifted storyteller and a shrewd businessman. And, of course, a family man. He represents the fourth generation to be working in the bergamot industry and runs the company that was established in 1880. The business has flourished while at the same time remaining within the family. Gianfranco's greatest success is having his twin sons, in their early thirties, work at his side: the fifth generation is already preparing to take over the reins. Most producers in Calabria and Sicily are also family-owned businesses, their names spanning the ages, their roots buried deep in tradition: Capua, Gatto, Corleone, Misitano, La Face . . . There are countless tales of lemons, mandarins, bergamot and even jasmine. For most buyers of natural ingredients, getting together with their Italian producers during the winter harvest is an unmissable event. Bergamot is the star of the show, and one must pay court to her.

Gianfranco and I know each other well. He speaks fluent French and knows how to transform himself at will into a character out of a Fellini film. He is capable of selling his latest harvest outright with nothing more than a skillful combination of a glance, a couple of words and a few waves of his hand. Whenever I ask him to recount the story of bergamot, he always starts in 1908. In that year, on December 28, the towns of Reggio and its Sicilian neighbor, Messina, were destroyed by one of the most violent earthquakes Europe had ever witnessed. At least 83,000 victims perished in the devastating quake and enormous tsunami that followed, an unimaginable toll. The region was utterly destroyed. Among the dead were Gianfranco's great-grandparents, who had established the business. It was an event that sent shockwaves through all Europe; nobody could have known that the continent would be rocked just six years later by a tragedy of an altogether different scale. A century later, in Reggio, the memory of that date is still somehow present; time passes slowly. The town's sleepy atmosphere is evident if one takes a stroll along its majestic seafront promenade, the longest in Italy and a beneficiary of the reconstruction. Many of the enormous ficus lining the esplanade are at least two hundred years old, poignant survivors of the tsunami. At the tip of the Italian boot, Reggio Calabria continues to look out toward Messina. The strait that separates the two cities is only three kilometers wide but they are linked by something considerably deeper than this stretch of water. As Gianfranco says, they will always be linked by the memory of that overwhelming tragedy 110 years ago. While elsewhere it may have been forgotten, in the memories of local families, including his own, it remains ever-present.

At the end of the esplanade, snowcapped Etna seems to emerge directly from the sea in the distance. The volcano is in Sicily, on the other shore, a symbol of the disjuncture between this unassuming coast and the illustrious island. Reggio still lags behind in terms of economic development and its lack of tourist infrastructure, always casting envious looks at the cruise boats berthing in Messina across the water, carrying thousands of tourists to the wonders of Taormina. But if Reggio slumbers, she does so with pride, for she knows her unique and indispensable role in the fragrance industry. Reggio is the bergamot capital.

The name "bergamot" is, thanks to tea, more recognizable than the curious fruit itself, which is similar in appearance to a lemon, its zest containing a unique essential oil. Fresh and potent, green, floral and zesty, bergamot oil is a treasure. The fruit is the product of the ancient graft of a species of lemon tree onto a bitter orange, the famous bitter bigarade or Seville orange tree, whose flowers are the source of the beautiful neroli essence, and whose fruit gives us bitter orange marmalade. The bergamot tree resembles lemon and orange trees, bearing fruit from December to February that is a paler, more subdued yellow than that of lemons. Its fruit can be either round or oval-shaped, growing in irregular size and shapes; there is an ill-disciplined, idiosyncratic side to the bergamot, heralding both the bitterness of its juice and the subtle perfume of its peel. It owes its existence, without a doubt, to the truly cult-like worship by the Arabs not only of orange trees but of citrus trees more generally. If we can agree on the fact that the orange tree originated in China, it was the conquering Arabs from the eighth, ninth and tenth centuries who brought the bitter orange all the way to Spain and, with it, countless types of lemons and mandarins, products of their accomplished grafting techniques. Theirs was a selection process focused primarily on the decorative and ornamental: perfumed flowers, the shape of the fruit, the burst of color against dark, evergreen foliage. The bitter orange remained a favorite in these selection processes because its blossom has the most refined scent and its hardiness made it conducive to planting not just in the gardens of palaces and mosques but as street trees in the towns of the Mediterranean south.

If grafting a lemon stem onto the stock of a bitter orange was an obvious thing to attempt, the result remained a secret for many years, probably on account of the bitterness of its flesh and the absence of brightly colored skin. This newcomer was given the Turkish name *Bey armudi* (the lord's pear). It was an event that took place in the world of perfume in the eighteenth century that led to bergamot's increased popularity. The year 1709 saw the development by a brilliant Italian named Giovanni Paolo Feminis of Aqua Mirabilis, which would in turn be refined by Jean-Marie Farina, who christened it "eau de Cologne," thereby launching a tale of extraordinary success that has endured for more than three centuries to the present day. Eau de Cologne was a revolution, marking the birth of modern perfumery. This elaborate blend of aromatic essential oils and alcohol launched the fashion for freshness and perfume when performing one's toilette. Napoleon adored it and his troops ensured it spread far and wide. At the heart of its formula, alongside the Provençal essential oils of thyme, rosemary and especially lavender, was essence of bergamot, which took center stage. Here, for the first time, over and above the richness of its own notes, it played the key role of enhancing the personality of the other essential oils. The success of eau de Cologne would lead to an ever-increasing demand for bergamot.

The first documented bergamot plantations in Calabria date back to 1750. Since then, they have remained confined to a narrow ribbon of coastal land starting a little north of Reggio and finishing at the same latitude, on the Ionian coast. It is as if the trees are reluctant to grow beyond that historic arc of land. Sicily is known for its lemons but bergamot does not grow well there. Nor have efforts to develop other sites in Ivory Coast and Argentina met with success, particularly when it comes to quality. Calabria is proud of its almost exclusive cornering of the bergamot market, and is doing its best to secure its future.

The fortunes of southern Italian citrus, which began with bergamot and the bitter orange, took a spectacular turn in 1850. These days we tend to forget the significance of the discovery in 1830 of the role played by the vitamins in lemons in the battle against the scourge of scurvy, which used to decimate ships' crews. It was a discovery that would transform the fate of sailors around the world and would facilitate new developments in maritime trade. The demand for lemons, particularly from American vessels, was such that within two decades Sicily was literally covered in lemon trees, its land being ideally suited to their cultivation. Having first grown them for the fresh fruit, agriculturalists now turned their attention to producing essential oil from their zest for the fragrance industry. From 1850 onward, and for almost a century, southern Italy would witness a golden age for its citrus essences that would mirror the boom in perfumery taking place in Grasse.

Calabrians are fond of telling this story, as well as recounting how, for more than a century, bergamot essence was produced entirely by hand using the renowned sponge and bamboo technique. Seated in front of a pile of cut fruit, the worker would rub the peel of half a bergamot against a bamboo stalk wedged up against a bowl in order to bring out the oil. In the other hand, he would hold a large piece of natural sponge that would absorb the liquid and which he would then squeeze out in order to collect it. In modern-day facilities, older workers still know how to use this timeless technique. One of them suggested I try it. With fingers covered in juice, and taking care not to lose the oil from the sponge, I copied my neighbor's actions, my nose overwhelmed by the fragrance released by the peel. Gianfranco's photographs dating back to the beginning of the twentieth century are quite something: fifty men and fifty women seated opposite each other in a great shed, lined up in a way that would not have looked out of place on one of Henry Ford's assembly lines, all pressing the peel against a multitude of absorbent sponges to catch the precious green liquid. A technical revolution would see the gradual implementation of a change in method. In the mid-nineteenth century, Nicola Barilla invented a machine to grate the fruit which was given the wonderful name of *la calabrese*: the Calabrian woman. With its cast-iron graters, its ingenious mechanical design and its chestnut case, the *calabrese* in time became the preferred implement for the production of essential oil and would help meet growing demand in the inter-war years.

Gianfranco has clear memories of that era, which are nonetheless a little colored by nostalgia. "You know, Calabria was a significant producer for the fragrance industry. In addition to bergamot and mandarins, there was plenty of top-quality jasmine. And from our orange blossom we also produced neroli, which we continued to do long after Grasse had given up on it. We had the advantage of being considerably poorer!" he says, smiling. Everything is different these days. There is no more neroli, and only a few acres of jasmine, which Giorgio, another essential oil producer, is determined to maintain in memory of his father. Always family. Southern Italy continues to produce essential oils from lemons, mandarins, bitter oranges, blood oranges and, of course, from bergamot, yet it is a challenge in the face of the citrus industry giants in South America and the United States.

Orange, lemon, lime, grapefruit: citrus essential oils make up the main raw materials used by the flavor and fragrance industry. Sweet orange essential oil produced in Florida and, more particularly, in Brazil, has the most significant market share by a considerable margin, and is a by-product of the orange juice industry.

After the fruit is pressed for its juice, the skins are then pricked or distilled to extract the essential oil. Manufacture of essential oils by this process bears no comparison to the specialist manufacturing method used in Calabria. Farms growing hundreds of thousands of hectares of orange trees produce fifty thousand tons of essential oils every year, five hundred times the amount of bergamot essential oil in Calabria! As for the lemon industry, the global market is dominated by Argentina, and Sicily's essential oil has become a niche product, relying on its quality for its survival. The competition is spread far and wide in the citrus industry: Mexico, South Africa, Turkey, India and China. The whole world is planting trees, keen for fresh fruit. Gianfranco the Calabrian is keeping a close eye on it all. He has considered and rejected opportunities to expand into Latin America. He is convinced that the Sicilian and Calabrian products have all the necessary attributes to justify persisting and continuing to expand there, the keys to success being quality, innovation and luxury. "Brazil is producing by-products of orange juice, whereas *I* am making perfume." And he is only half joking. "We have exceptional fruit, a long tradition of cultivating our land and the expertise that goes along with that. And we have the support of the world's best perfumers. Every winter they come to smell the new bergamot and mandarin harvests because they know they are unique."

For a dozen or so years, we have been witnessing a veritable return to favor of Italian essential oils. The mobilization of Calabrian and Sicilian producers in their response to the fragrance industry's present-day requirements has paid off. Using sophisticated equipment, they are distilling ever more refined products to service every sector of their client base, from brands of carbonated drinks to top-of-the-range perfumers. It was high time to renew the faith of purchasers in a bergamot product that had started to receive a bad press. Quantity was as unreliable as the price, and the quality—often recomposed blends, removed from the pure product—left a lot to be desired. The spread of mediocre oils had ended up undermining bergamot's appeal. But between the efforts of farmers to improve their crops and those of producers to ensure quality control, the industry has started to turn things around, bringing a smile back to the faces of those in the profession.

It is the end of the season, and we are off to see the harvest around San Carlo, a small village set back from the coast, at the southernmost tip of the peninsula. As soon as one leaves the town of Reggio behind, the countryside is covered in citrus orchards, appealing in their diversity, so much a reflection of the region. Below the motorway there is a passing parade of lemon trees in the courtyards of houses, small groves of mandarin trees, and fields of bergamot in every shape and size, young trees and old, closely trimmed and planted in organized rows, or too tall because their aging owner can no longer reach to prune them. We follow the narrow road that climbs up through the village, passing trailer-loads of fruit. Crates of yellow spheres are stacked alongside the fields, fruit that has been picked by families or with the help of a team of laborers, Calabrian or immigrant—either is possible. It is a forty-kilometer trip for the trailers delivering the fruit to the factory. Everything looks typically Mediterranean: washing at the windows, small plots climbing up the hill only to stop short at the foot of the next mountain, donkeys and goats placidly watching on, yellow splashes of wild mustard, the brilliant green of foliage against the blue of the sky. I scratch a piece of fruit with my nail; its skin releases a burst of fresh green notes, rich and addictive. Once you have had a chance to smell fresh bergamot, it will stay with you forever.

Most of the farmers around San Carlo belong to the Consorzio, the big cooperative established in the late 1930s to revive production. It has survived until now and, at Gianfranco's instigation, has become one of the driving forces behind the industry's revitalization. Farmers have planted twelve hundred hectares of bergamot and the yield is so plentiful that the cooperative is scoping out more than four hundred hectares for new plantings.

Gianfranco's facility is located just outside Reggio, housed within the Consorzio's historic buildings, where essential oils continued to be hand manufactured in the pre-war years. In these great warehouses, a sea of fruit slowly makes its way along conveyor belts toward stainless-steel machines with mysterious names such as the *pellatrice* or the *sfumatrice.* The fruit is pricked, scrubbed, split and pressed, with juice and essential oils separated out. The first oils are then put into a centrifuge to remove any water, before the decanting and final filtration. The scent of delicate green bergamot essence fills the laboratory. Every batch is analyzed before it can be blended with previous batches. Giandomenico and Rocco, twins and representatives of the fifth generation, give me a tour of the facility's latest innovations. They have inherited their father's charm and passion, and there is talk of traceability, security and technical progress, catchphrases of today's industry. Investment and modernity. They are proud to show me the gleaming, brand-new centrifuges, their performance five times more effective than the model installed by their father. They are outgrowing the facility, the business will have to move to a new site looking out over the sea where it will double capacity.

The tranquility of the area, the prestige of its essential oil and the producers' expertise are all factors drawing in the visitors. Traditionally, sourcing agents in the industry would come to Calabria to negotiate with Gianfranco and his competitors. But for some years now, perfumers, marketing directors and journalists on the hunt for beautiful images have also been flocking to the area. The producers' transparency and the manner in which they are industry standard-setters also mean they feature in profiles on sustainable development.

Back in Reggio, after our wanderings through the ancient mosaic of seaside orchards, we start our negotiations with Gianfranco and one of his sons. We are in agreement regarding the bergamot, but there is animated debate around the lemon. Father and son are trying to persuade me of the exceptional quality of essential oils that we are not as yet buying from them. A purchaser is watchful, an onlooker is able to be seduced. Both men talk animatedly, echoing each other's arguments in an attempt to convince me of their opinions. We smell, we scratch the fruit, we go over the lab reports, they stand up and sit back down, their gestures a call and response. Passionate and convincing, by turn farmers and chemists, to me they are the true heirs to the noble story of eau de Cologne. I leave with samples; things are on the right track.

---

That evening, Gianfranco and I take the ferry to have dinner across the water, in Taormina. He tells me of the planned Messina bridge, which prompts talk of Homer and the *Odyssey.* Twenty years earlier, I had not wanted to leave this Mediterranean heartland without my own reckoning with Odysseus. I had headed to the mouth of the strait, just north of Reggio, to see the rocky outcrop that looks out over the village of Scilla, unchanged, I imagined, since it made its terrifying appearance in the *Odyssey.* Sure enough, the cavern which sheltered the six-headed monster is there in the cliff. By having Odysseus falter between Charybdis and Scylla, Homer skillfully transformed into myth the very real threats that locals and seafarers have had to confront since the dawn of time. This arm of water is so narrow and so deep that sailors are always faced with challenging whirlpools on the Sicilian side and powerful currents off Scilla on the Calabrian coast.

The notion of a bridge that would link Calabria to Sicily is appealing on a map; it has had as many supporters as it has had detractors. Grand plans have been drawn up for a suspension bridge, the only feasible technical solution given the depth of the water. With a span of more than three kilometers, it would be the longest suspended roadbed in the world. The bridge was almost built ten years ago, before plans were abandoned for political and financial reasons, Gianfranco tells me with a sigh that says it all.

On either side of this unbridged strait, the Sicilians and Calabrians enjoy a rivalry and a certain solidarity. Freightloads of lemons and bergamot cross paths on the ferry; graters and presses turn in factories on both shores. On the Calabrian side, people are keen to continue planting, picking and pressing their bergamot. Here, and nowhere else. Reggio will preserve the shade of its great ficus trees, its history and its orchards, quietly confident that ever-increasing numbers of tourists disembarking in Messina will want to make the crossing to discover the beauty of Calabria.

I experienced an astonishing array of feelings on the short crossing from Reggio to Messina. Images of Odysseus on his ship, of the 1908 earthquake, of the bergamot sponge workshops. An overlapping of fragments from the tales of which Gianfranco was both custodian and narrator. Sitting across from this man for whom bergamot is part of his very essence, how I wanted to be able to imagine that bridge, liberated from Homer's monsters, finally overcoming old wounds and uniting the two towns, bringing together the people, their orchards and their fruit in the promise of a shared future.

# THE MASTER AND THE WHITE FLOWER

*Jasmine, from Grasse to Egypt*

"I want to make the best naturals in the world, and for that I need the best sourcing agent in the industry. And that's you." Jacques had tossed the comment my way one day in 2009, looking me right in the eye, as direct as he was charming. He was suggesting I join him at the large perfumery company where he himself held the position of master perfumer. I had been in the Landes region for more than twenty years, and ten minutes was all it took for me to know I would accept his offer. A star in the industry, his title amounted to a prestigious calling card, awarded by perfume houses to a chosen handful. It rewarded a career marked by significant successes for perfume brands such as Issey Miyake, Jean Paul Gautier and Stella McCartney. Master perfumers are the aristocracy of fine fragrances, responsible for creating the most famous perfumes of the last two or three decades. They represent the pinnacle of a profession that blends art, artisanal skill and relentless hard work. A profession incorporating inspiration and the irrational, passion and a touch of sorcery. Jacques and I had known each other for ten years. As producer and buyer I offered him the highest-quality products I could find. Opportunities were rare, and it was essential to grasp them and be at one's persuasive best. He would quickly identify the most beautiful samples, was precise in his judgments, and pulled no punches in his assessments, always on the hunt for an olfactory facet which he would know how to make his own. Warm and curious, Jacques would have me tell him about my travels, stories that he would regularly punctuate with the comment "I absolutely have to see that!" He loved to hear stories about the origins of raw materials. We got along well. A fanatic for natural ingredients, and armed with his reputation and his charisma, he had persuaded the company to let him run a research and development laboratory that they were establishing in his home town of Grasse. We worked together for more than three years and from time to time would set off to find the source of the products he loved. Jacques was particularly drawn to jasmine; we had been on the hunt for it for years, and from my point of view, it was a constant journey of discovery. To follow a natural ingredient back to the fields, the facilities, into the hands of the pickers, to witness the emotions and choices of a great perfumer, these are the unique experiences that make my work meaningful. I have touched, picked and smelled jasmine while listening to his stories, I have been carried away by his passion, his memories and his firmly held opinions, witness

*Mehalla, Egypt, the end of a morning picking jasmine flowers at Sayed's farm*

to the paths he has taken on the way to the secret garden of his formulae.

For me, the smell of jasmine embodies something approaching absolute beauty. As it finds its way to our brain, the scent of its blossom triggers an immediate sensation of happiness. Heady and beguiling, both familiar and foreign, jasmine unsettles us by evoking the sweet mildness of Mediterranean gardens mixed with exotic, intoxicating, almost animalic aromas. In the world of perfume, it has long been associated with Grasse, the town that still aspires to be jasmine capital of the world. With the intense blue of the Provençal lavender at its back, Grasse looks out in the other direction toward the Mediterranean and to the fragile white of the jasmine blossom.

From its perch above Grasse and Cannes, the village of Cabris offers on a clear day a splendid view out to Corsica. Jacques has both his roots and his home in Cabris; it remains his point of reference. His family has lived there for generations, his great-grandfather was the mayor, and his grandfather and father preceded him in the perfume industry.

In the summer of 2010, as we wander through his garden, he points out his beds of roses and tuberoses between the olive trees and, running side by side, a row of lavender next to a beautiful row of jasmine. He picks some flowers, delicate five-petaled white stars, gathers them in his palm and brings them to his nose as he closes his eyes. A moment of silence. He tells me to inhale their scent and says softly, "I would love to go on a hunt for some exotic jasmine with you. But you know, nowhere else could it smell as good as this. The jasmine here is beyond compare." His dark eyes crinkle and he looks at me with the impish smile he has managed to retain. "Smell that volume, that depth, that richness! It is green, animalic, miraculously balanced. And what's more, it enhances the other natural ingredients in the formulae." For Jacques, jasmine symbolizes the pre-eminence of the Grasse terroir in perfumery. That afternoon, he invites me into his universe, a blend of family culture and personal experience. He tells me its history as if he has been living it himself for more than a century. As he talks about the jasmine, he talks of his family, of his admiration, his gratitude. I, too, can sense his emotion.

The *Jasminum grandiflorum* species of jasmine with its large blossoms comes from northern India, and was brought by the Arabs to Spain, Italy and France in the 1650s. Adopted throughout the Mediterranean region, its success in Grasse was rapid, as evidenced by the existence of fifteen hectares of gardens dating back to the seventeenth century. When, in 1860, the perfume houses started to establish themselves in the colonies, the building of the Siagne canal permitted the irrigation of hundreds of hectares, and jasmine started to take off. Two hundred tons of flowers were harvested in 1900, six hundred in 1905. It reached its peak in 1930 with a harvest of 1,800 tons, a phenomenal volume. In the harvest season, an army of five or six thousand pickers working from July to October would pick twenty thousand flowers each day. Work would start between four and five o'clock in the morning because the blossoms open overnight and the best jasmine is harvested before it has even seen the sun. It was labor so typical of rural life in those days, for the most part arduous but joyful at times, too, as borne out by the remarks of one picker. "The harvest was often carried out by Italians. Calabrian families with their children, who of course would also work. They were put up in cabins and the farmers would provide them with vegetables. Among the jasmine, the atmosphere was very cheerful: the pickers would sing, others would join in, sometimes we would even be treated to a solo or duet. The Italians had so much spirit, they were much loved."*

Jasmine does not lend itself to distillation—its yield of essential oil is too meager—rather, its unique scent must be captured by "extraction." For a long time this was carried out by the ancient method of *enfleurage*. A layer of fat would be smeared over a plate of glass called a chassis, onto which the flowers were spread. They would be left there for a day or two until the fat had soaked up the scent. The fat was then scraped off so it could be washed down with alcohol until a concentrate known as an "absolute" was obtained. These delicate tasks would be performed by hundreds of women who were the aristocrats in the labor hierarchy of the factories.

Jasmine absolute very quickly became one of the flagship products of perfumery, with demand for its olfactory splendor growing continuously until the 1950s. At the end of the nineteenth century, the fat started to be replaced by more effective solvents such as benzene or hexane, nowadays the most common way to produce jasmine extract.

This current method does not have the magic or beauty of the *enfleurage* process, and Jacques would grow animated when reminiscing about the traditional procedure. "There was something

* *Souvenirs* by Simone Righetti, September 2005.

fantastical about it. How I would love to see the process brought back. We need to retain the principle but modernize it."

The story of jasmine is an inextricable part of the history of the perfume industry in Grasse. The harvest record set in 1930 would mark the peak of the use of natural ingredients in the fine fragrance industry. Local businesses were powerful and well known, planting extensive crops of flowers around the town and throughout the region. Flowering orange trees covered the terraced hills in the area through to Vence and Bar-sur-Loup and producers set up large lavender distilleries in Haute Provence. Orange blossom, rose, lavender and jasmine were the heroes of this great saga that spanned two centuries. In the second half of the nineteenth century, perfumery took off once again: several businesses followed the colonial conquests of the French army, looking to establish facilities in those countries where raw materials might be grown, especially in the tropics. Jasmine was to be part of this story, and the expansion of these new horizons would take it far from Grasse, first southward and then further to the east.

The company trading under the name of Chiris was the jewel in the crown of this story. For seventy years, Léon Chiris, followed by his son Georges, built a worldwide network of trading posts, factories and plantations that left a considerable legacy. The extent of their business interests was impressive: large plantations and extraction facilities in Boufarik in Algeria; production plants in Guyana, Madagascar, the Comoros and what was then known as the Congo and Indochina; and businesses in northern Italy,

Calabria, Bulgaria and even China. Chiris was hungry for spices, essential oils and extracts; he was probably the first to have understood the significance of planting, collecting and distilling at source to produce the best perfumes. Rosewood, ylang-ylang, vetiver, benzoin, geranium, vanilla, citronella, musk: all of these ingredients converged in Grasse. From the 1930s onward, jasmine was cultivated and extracted in Calabria, Morocco and Algeria, significant additions to the company's domestic production capacity. Today, only a few hectares from that era remain in Morocco, and in a single field in Calabria; there has been no Algerian production since the end of the 1970s. The divine white flower made its way to more distant shores, first to Egypt and then, some thirty years later, to India.

A fascinating book entitled *L'Âge d'or de la parfumerie à Grasse** recreates a little of the Chiris saga in photographs. In some of the images, the smell of dust and sweat is more evident than the scent of perfume, as in one particular photo showing Congolese workers carrying a sedan chair that contains a foreman in a pith helmet through the fields of citronella, a crude reminder of the foundations on which the colonial world was built.

At the Paris Colonial Exhibition of 1931, Georges Chiris played lord and master, hobnobbing with heads of state and receiving political support from the highest levels.

The combined effects of the crash of 1929, the Second World War and the demise of the French colonial empire dealt a fatal blow to Grasse's prosperity. The local industry would never truly recover.

* Éliane Perrin, Édisud, 1987.

---

I was first introduced to jasmine in the plantations and workshops of Egypt and India, where I bought extracts that Jacques would use in his compositions. One September, we headed off together to the Nile Delta to see the jasmine that grows on the far shore of the Mediterranean.

Sayed is one of the three largest jasmine producers in Egypt. A man of mature years with a stately demeanor and an intensity to his gaze, this mechanical engineer is also a professor at Cairo University, and passionate about both his country's history and its future. Warm yet shy, determined and curious, Sayed has a great deal in common with Jacques, including a big personality. The producer and perfumer had already crossed paths in Grasse where they had enjoyed each other's company, and Sayed had asked me to persuade Jacques to pay him a visit.

After a few attempts to set up a facility in Egypt prior to the war, the first local extraction plant commenced operations in about 1950. But after Nasser came to power in 1954, the nationalizations and agrarian reforms led to the dismantling of the plantations and brought work in the factories to a halt. When 1970 saw another change in policy, Sayed's father was encouraged to invest in a new facility and to plant some jasmine. He was part of the state's plans to revitalize the economy, and the great push to mobilize Egyptian exports, which would be used to pay for arms supplied by the U.S.S.R. Jasmine concretes joined locally made furniture and shoes in an unlikely and astonishing swap: flowers in return for weapons! The basis of any number of illicit deals, these exchanges even led to poor-quality Soviet soap being perfumed with real jasmine, a story which some thirty years later still sparked hilarity on the part of our Egyptian friend. Were the Russians even aware of this incongruous luxury?

Three hours from Cairo, in the heart of the delta, the countryside is unrelentingly flat. Villages of unfinished brick houses are surrounded by fields and canals, the blessing of the Nile waters. There are farmers everywhere, men in white, women in colorful dress, gaggles of children accompanying their parents in the fields or playing with balls on hummocks of uncultivated land. No matter where you are in the world, there will be children playing with balls, no matter how improvised the pitch, no matter the state of the ball, be it inflated or flat, made of rags, or nothing more than a sphere of knotted plastic bags. From Madagascar to Guatemala, from Haiti to Morocco, there are always children laughing and playing.

The yellow of the stubble fields stands out against the black of the soil. We find Sayed's farm at the end of a row of palm trees running along the fields of geranium and orange trees. At the center of his property stands a large red house with a roof terrace that offers a view over the surrounding crops. The fields of jasmine bushes are a marvel, luxuriant and literally covered in blossom. The legendary fertility of the soil, a product of the ready availability of water from the Nile, coupled with sunshine, yields spectacular results. Large flowers, record per-hectare yield, and a very beautiful product: an absolute that is almost sensual, with sunny, fruity, deep, gourmand notes. It is nine o'clock in the morning and already it is very hot. Dozens of women of all ages who have been picking for the past four hours are starting to bring in their baskets to the weighing station. The excitement is palpable; news of the arrival of a famous perfumer has made its way around the farm. Sayed also grows and extracts geranium and violet leaf, and has started distilling neroli, the essential oil from his bitter-orange trees. His visitor is keen to see operations in progress and to put his nose in everything, and also to understand the impact of the time of day when the flowers are picked on the quality of the jasmine absolute. He wants to bring back a work schedule for his teams in Grasse, and to find new ideas that will enrich the perfumers' palettes. Like princes, we have settled ourselves in handsome cane armchairs around the flower weighing station at the edge of a big field. Jacques peers into all the baskets, burying his head in them. The women laugh. The queue for the scales is an animated, cheerful sight, the women's clothes an explosion of bright colors, their baskets overflowing with jasmine. Sayed issues instructions in a loud voice, a pharaoh in his fields. The addictive white flower is having its effect, Jacques is bedazzled, brimming with enthusiasm. "You do see, don't you, Egypt is the Grasse of fifty years ago with its jasmine, its geranium, its orange blossom and violet leaf. Perhaps I should think about setting up a farm here . . ." he says to me a little later, thoughtfully.

That evening, back in Cairo, Sayed invites us to share a hookah on the banks of the Nile. Surrounded by fragrant smoke, he shares his own uncompromising views on the huge challenges involved in educating the youth, the missed economic and political opportunities, his assessment of the need for a pharaoh to lead this country. He is not short on arguments to rebut our Western thoughts and reasoning. I like his intelligence, his frank opinions. I take another puff of the shisha.

After three days together, we are talking about the future and of a long-term partnership. For a few hours, the man from Grasse has become an Egyptian. "So, Jacques, true jasmine, is it to be found in France or in Egypt?"

He feigns dismay at my question. "We need them both, of course! Finesse and gourmandise. A subtle marriage but one which will lead to wondrous things, unique things, provided we know how to work with them!"

Before leaving us, Sayed asks me for news of the jasmine industry in India, the other large supplier of the market, and a formidable rival. I know he is anxious. He has witnessed the arrival of the Indians onto the production scene, the progress made in their extracts, their increasing competitiveness. Output in both Egypt and India is now comparable, even if the quality is different, and most buyers source supply from both countries. I don't have the heart to tell him that I'm about to head off there as I'm considering a partnership with our Indian counterparts. It is not something that will threaten our business with Sayed but it could mean we prioritize our expansion in India. It is never easy to manage the rivalry between producers of the same floral extracts in different countries. Rationality should win out over sentimentality, of course, but when dealing with natural materials, the human element is never far from the decision-making and strategizing.

As it turned out, Jacques did not have time to plead his case for a farm in Egypt or to accompany me to India. A few months after our trip, he went to meet with the world's biggest luxury group. In a poor attempt to hide his pride, he confided in me that the owner of the group had greeted him with a pitch that came as no surprise to me. "I want to make the best perfumes in the world, and for that I need the best perfumer in the industry. And that's you!" The same words that had persuaded me three years earlier . . . We both smiled at the thought, even if this time the positions were reversed! A fierce defender of natural ingredients, who prized excellence and simplicity in perfumes, Jacques would have complete creative freedom to work with the most beautiful raw materials. Once again we would be working in different companies, but that was no reason for us to stop seeing each other. He would continue to need me to accompany him to the source, helping him find both the products and the stories of their producers. We very much intended to continue our shared journey.

# THE ELEPHANT AND THE WEDDING

## *Jasmine in India*

Madurai, a city in the southern Indian state of Tamil Nadu, is the nation's flower-growing capital. Whenever I visit, I never miss the chance to stop by the Temple of Meenakshi to offer up my head to be stroked by the trunk of the temple's elephant guardian. The warm, moist blessing provided by the beautiful creature decorated with flowers and paint must be accompanied by a wish. As I approach Malachi in the winter of 2011, I know what my wish will be. In a country that celebrates weddings with an avalanche of flowers, I am keen to bring about a marriage between the jasmine growers and the company I work for. It would be the culmination of my longstanding relationship with Raja and Vasanth, two talented businessmen.

The temple is one of the most famous in India, a city within

*Left: Raja, enjoying the smell of jasmine on his farm in Coimbatore, India*

*Below: A jasmine picker with both hands busy*

the city, and I love wandering barefoot across its warm flagstones, transported each time by the hustle and bustle and the piety of the crowd. There are flowers everywhere, stalls of garlands in every size and color, hanging from the necks of pilgrims and on the altars where the *agarbattis,* or incense sticks, are burning. The temple is an immense labyrinth, an unforgettable immersion into Tamil culture, a hive of activity where every day tens of thousands of people make their way through a maze of sculpted stone galleries before stopping for some private reflection at a golden column or a branch embedded in the ground, a relic of a sandalwood tree said to date from the time when the temple was first built. At the heart of the temple, open only to Hindus, a human-scale statue of Meenakshi carved from a single block of emerald takes center stage.

Flowers are grown in villages right across southern India's flower belt before being sold at market, woven into garlands and exported throughout Asia. The queen of the flowers in this province is a species of jasmine. Not the jasmine of Grasse nor that of Egypt, but a tropical jasmine known as *sambac* jasmine. Grown here for more than two thousand years, it is southern India's most popular flower, an ever-present facet of life here. It is the flower worn by women in their hair every day, the flower that is hung next to Ganesh, the elephant God, on the rear-view mirror of cars, the flower you will see as offerings in every temple. *Sambac* jasmine is the most common flower in the sumptuous garlands used for celebrations, the flower of weddings. It is fleshier, less fragile, more tropical than the *grandiflorum.* Its scent is intense, less animalic and sweet, with notes of jam or bonbons.

To witness the harvest in the fields surrounding Madurai, one has to leave the city at daybreak. Small roads pass through villages, a mix of bare bricks and houses painted in varying shades of indigo blue, a beautiful limewash that is diluted according to the means of each family. Every village specializes in growing a particular type of flower. There are carnation villages, and tuberose ones, but *sambac* is grown everywhere. While income from growing flowers may well set some Indian villages apart, they nonetheless remain impoverished: no sanitation, a hand-pumped water supply, undernourished cows and goats, children everywhere. The Indian countryside in these parts is an explosion of color: small fields of white, orange or red flowers, and plots of vegetables in every shade of green, vivid against the ocher-colored earth. An old man in turban and loincloth is plowing the fields with his two oxen and wooden plow, with not so much as a metal plowshare. How long ago was it that a wooden plow was used in France? Here, it is being used to prepare the soil for the planting of flower seedlings that will be used in our perfumes. Nothing seems as far removed from a flacon in a perfumery than this farmer, and yet, without his knowing, he has played a part in its preparation.

Women of every age are harvesting flower buds, a magnificent sight in the dazzling colors of their saris, a little garland of *sambac* in their hair. The buds will not open until twenty-four hours later, in the time it takes to transport them and sell them at market. Women work the rows, a few men accompany them, children are at school. The flowers are sent off to market around mid-morning; the first deliveries will attract the best price. The large marketplace in Madurai has retained its traditional feel with dozens of small stalls. From here, fresh flowers are sent to every big city in the country and as far afield as Dubai and Singapore, or even to Europe. Lively transactions are going on around the containers of flowers and in the garland stalls, where multicolored floral necklaces are expertly crafted, some of which weigh too much to be worn comfortably around the neck. They are threaded up by men who squat on their heels in their stalls, knotting each individual flower onto a string with a speed and dexterity that defy belief. The garlands are a riot of color, a blend of the white of the tuberose and *sambac* blossoms with the yellow or orange of chrysanthemums and carnations, the red of cockscomb or roses and the pale green of *davana*, the local wormwood which has such a lovely scent. The smells of the market are no less an assault on the senses. Food carts, rotting flower waste, puddles of stagnant water, exhaust fumes from hundreds of scooters . . . the perfume industry seems a world away from this hustle and bustle and the frenzy of colors and plants. Yet the use of *sambac* is ever more popular among perfumers. Its gourmand, white floral notes appeal to many people and competes with traditional jasmine.

Cultivation of Mediterranean *Jasminum grandiflorum* or Spanish jasmine was for a long time limited in India. It is less sought-after at the markets than *sambac* jasmine because its flower, lighter and more fragile, is less suited to garland-making. Things changed at the end of the 1970s, when the fragrance industry saw an opportunity to grow and extract the essence of its benchmark jasmine at its source in India and at low cost. It took the Indians less than twenty years to catch up with Egypt and become a producer to be reckoned with.

---

When I made that trip in 2011, I had already known Raja and Vasanth for fifteen years; they had set up a business which was now well and truly recognized in the industry as producing the best extracts of jasmine and other Indian flowers. At the start of the 1990s, these cousins, originally from Chennai (formerly Madras), were still students in Great Britain and the United States when their family inherited a flower extraction facility by way of repayment of a debt. From one day to the next, with no time even to reflect, they found themselves producers of jasmine for the world of perfumery. The business became a success story, to which Raja lent his sales acumen and charisma and Vasanth his nose for finance and strategy. Their company played a key role in shoring up Indian jasmine's reputation on the international market as a credible and reliable product, patiently establishing two fine factories, one in Coimbatore, in the Spanish jasmine region, and the other close to Madurai, in the heart of the area that produced *sambac*. Within a few years, Raja and Vasanth had seduced the industry with a combination of respect for the farmers and their knowledge of the flowers, the quality of their products, and a new and reassuring image of a country that did not have a history of inspiring confidence in buyers.

For more than a century, Firmenich, the Swiss company I had joined in 2009, had built its reputation on chemistry and the creation of innovative aromatic molecules, a culture at some remove from fields of flowers and their distillation. At Jacques's encouragement, Firmenich had acquired a company based in Grasse and found itself plunged into the world of natural raw materials, and on the hunt for a strategy. I had been engaged to look after the naturals side of the business and on my arrival my opinion was sought as to what forms of investment I would recommend. I advocated a model that would involve partnering with the best producers of essential oils and extracts that already existed around the world. Acquiring a financial interest would provide a partner with our technical and financial support, long-term guaranteed sales and collaborative investment in innovation. My first suggestion was to invest in Raja and Vasanth's company. The wealth of its aromatic products combined with the capability of these partners made India an ideal candidate for a first attempt at marriage.

I had not anticipated the difficulties we would face. Within the business, opinions were very divided as to the chances of a successful marriage with an Indian partner. Raja and Vasanth also had to be persuaded to take the plunge. This union of businesses as different as oil and water was as unnerving as it was appealing. Vasanth knew that his company would be able to expand much more quickly with such backing, but the two cousins feared being swallowed up. It would be necessary for those who shared my belief in the project to be patient and persuasive in order to overcome any skepticism and reluctance.

After three years of negotiations, Patrick Firmenich, the company's chairman, wagered that, regardless of their size, one very large, one very small, the two companies were nonetheless both family-owned, sharing similar values. I have a vivid memory of the meeting in Geneva between Patrick, Raja and Vasanth just after our agreement was signed in 2014; there was a palpable emotion in the exchange of speeches and compliments around the decision to merge interests. I was profoundly happy on the occasion of the marriage, as I saw in it a model for the future of raw materials: it amounted to acknowledgment by a large fragrance company of the importance of those working the fields and distilling the essences.

In the autumn of 2014, I came to Madurai to get an understanding of what would, from this point onward, be a joint endeavor, a form of celebration. The company had already become the market benchmark for the two Indian jasmines. In Coimbatore, they were sourcing Spanish jasmine from a network of a thousand small producers who were growing exclusively for them. In Madurai, *sambac* was purchased at the flower markets at day's end, when prices dropped because the flowers were too blown. Wedding flowers cost considerably more than fragrance flowers, but demand from the extraction units provides the growers with secure sales. Back at the factory, in the early evening, Raja, Vasanth and I observed in silence the heavy carpet of *sambac* flowers spread out over the great cement flagstones, a handsome illustration of their success and our alliance. Twelve hours after they had been picked, they had almost opened and could now be processed in this facility that operated day and night, seven days a week. Together we were enjoying the success of our marriage. A site tour always includes a visit to the tree planted by each visitor on their first visit. On the day I was visiting in 2014, my tree was already ten years old. The tree was an Indian cork oak, also known as the jasmine tree, because of its profusion of white, scented blossom. It had become a reference point for me and an accomplice over the course of my regular visits. In the late afternoon, Raja accompanied me to the Temple of Meenakshi. He knew I absolutely had to see Malachi. As I approached the elephant, I looked her straight in the eyes, she nodded her beautiful head, gently flapping her ears, and then, surrounded by the scents of the temple, I sought the caress of her trunk. She left it there a little longer than usual, as if to tell me that she knew my wish had come true.

Drawn to the country, the flowers and our partnership, perfumers are coming to visit Raja in increasing numbers. He knows everyone in the world of fine fragrances, from New York to Paris, attracting clients with his charm and competence. Over the years, he has become a star in the small world of perfumery. With his patience, conviction and elegance, he, along with his cousin, represents a new generation of Indian entrepreneurs who are talented and perfectly in tune with the Western world without in any way forgetting their roots.

For a long time *sambac* jasmine extract was an exclusively Indian product, until China surprised the fragrance industry by starting to supply its own local *sambac* samples. Raja was aware of what was happening and had spoken to me about it. The first samples were poor, but fairly quickly a decent-quality product began to appear. Two years after his departure, I continued to see Jacques on a regular basis. Always keen to hear about new products, he would often ask me for news from the jasmine world, in Egypt and India. When I showed him an extract of Chinese *sambac* he was impressed and murmured, "What if the two of us were to head off on a little jaunt together?" So two months later, there we were, off to the far reaches of Guangxi province in southern China.

---

A day's drive from the provincial capital of Guilin, the fields around the city of Heng Xian are where most of the country's *sambac* is grown. Having visited the fields of the Nile Delta, Jacques and I now found ourselves wandering through a large jasmine crop in a landscape criss-crossed by high-voltage power lines and their pylons, with a view over an abandoned factory, a not unusual vista in the Chinese countryside. I was a regular visitor to China; it is a source of important essential oils for the fragrance industry, such as eucalyptus and geranium. The many hundreds of rural distilleries and their operators always remind me of the harsh nature of the work, as if it were a job one did for want of any better option, a last resort before finally managing to find work in the city. There is no real place here for tradition; previously one would work for the state, now there is another boss. The only thing that matters is the price of the product. Chinese *sambac* jasmine has been planted in great quantities on well-maintained terraces. Flowers from these fields are used to perfume top-of-the-range brands of the famous Chinese jasmine tea, a significant market that justifies the careful cultivation of these crops. Silent and swift, a small team of pickers are hard at work, bamboo hats on their heads, bags of flowers strapped across their bodies. Jacques is deep in conversation with one of the pickers, a southern Chinese woman who wears a blue jacket and large hat and whose face is etched with lines. I am able to make out that they are discussing the technique of plucking the flowers. Grasse may well be a long way away, but Jacques is right at home. Much like the *sambac* in India, these farmers' flowers are destined for the flower markets, where they will first be purchased by tea producers. Great blankets of green or black tea are spread out in their factories in vast, hangar-like spaces, ringed by a belt of fresh flowers. The spectacle of this geometric arrangement verges on the ceremonial, the black of the tea contrasting with the white of the flowers, and the powerful scent of the two distinct fragrances before they are blended adding to the atmosphere. Left to infuse the tea for a day or two, the flowers slowly release their scent in a sort of basic but very effective *enfleurage*. When the flowers are subsequently gathered back up and sorted, their scent will have diminished significantly.

Our host in Heng Xian is my local supplier, Jack. A wine enthusiast and bon vivant, Jack is delighted to have us visit and is conscious that my friend's presence is somewhat unusual. Jack and Jacques hit it off instantly, and I am amused at the French perfumer's effusive praise for the Chinese red wine, which is flowing like water. However, it is Jack's *sambac* concrete that Jacques really likes, even more than his wine, preferring it to Raja's product. For a long time, Chinese products were considered poor quality, because producers would use flowers that had already been used to infuse tea, but since Jack has started making concretes from fresh flowers, his products have won unanimous recognition. Back in the town's flower markets, we wander through a multitude of mountainous white piles, *sambac* buds and dainty corollas of magnolia, the other main local flower. We buy a bag of *sambac* to sample, spread out the buds on my hotel room floor, allow them to open overnight, then smell the open flowers in the early morning. I fall asleep with their green, sugary scent in my nose. The next morning, Jacques smells the flowers before leaving the hotel and murmurs, "No comparison with the buds. Magnificent." There is something about jasmine that always plays out between night and day; it prefers to open up away from the light, then show itself off at first light.

Two years after our trip to China, Jacques's first collections were launched in his label's boutiques. Perfumes rich with far higher proportions of naturals than usual. Both Egyptian jasmine and Chinese *sambac* featured in these compendiums of the most beautiful ingredients. And true to his word, he also used jasmine from Grasse. These days only twenty-odd tons of jasmine flowers are harvested in Grasse every year, barely more than one percent of the 1930 harvest, but new crops of roses and jasmine have recently made an appearance, symbols of the perfume capital's revitalization and of the "return to grace" of its flowers. The most prestigious labels are once again aware of the exceptional character of Grasse's legendary terroir and they are keen for it to make a comeback, whether real or symbolic, in their compositions and in their marketing, especially.

Jacques has long been a devotee of flowers from Grasse, but he knows that the price of those jasmine flowers is forty times higher than those sourced in Egypt or India. He is convinced that the future of Grasse jasmine will be determined by new extraction methods capable of even more faithfully reproducing the scent of the flower blossoming on its bush, and thus producing something incomparable. We have started working toward this and the results have, he says, provided him with a truly unique product. It is remarkable that perfumers and their labels are once again eager to use jasmine from Grasse. It is particularly impressive given the utterly extraordinary prices that extracts from these blossoms command. Does there exist, then, a perfume house that is willing to seek out the extraordinary when it comes to raw materials? Many perfumers dream of finding something exceptional, even if the industry only provides a tiny number of them with the means to make use of it. And yet that dream is a need, it is the very essence of perfume.

Lingering in the delicate scent of these flowers is the mystery of the creative process itself. I had spent a long time observing Jacques on our jasmine wanderings. His excitement was spontaneous, then considered; a passion for one product would disguise another. He did not hold back when he was with Sayed, Jack or me, but I realized that some of his remarks were for his own benefit, like pieces of a creative puzzle that only he, the master perfumer, could solve. A particular scent of jasmine evoked a dozen other naturals which he would then start to compose in his mind. "When I was a child," he had confided in me, "there were still many jasmine fields around us and, like every person from Grasse, I used to love smelling its scent on the breeze in the morning and evening. When I was very little, I used to love to gather up some of the flowers from my father's factory. He would smell them with me back at home and would put them under my pillow in the evening. I have a very precise memory of that scent. It was the start of my training as a perfumer . . . allowing myself to be carried away by my emotions and realizing that you can commit them to memory. Later on, you learn how to classify them, every ingredient has its own emotion. Every creation is a mosaic of emotions." I have loved following him on his olfactory wanderings, a path that started with the flowers placed under his childhood pillow. When Jacques gives free rein to his sense of humor, his face lights up with the smile of a little boy. That same little boy who was lulled to sleep by the words of his father, carrying the scent of jasmine into his dreams in the land of master perfumers.

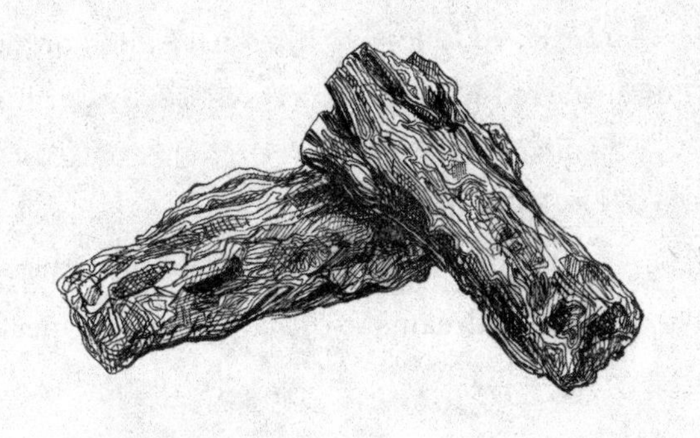

# THE PIONEER AND THE TREE TAPPERS

*Benzoin from Laos*

"Monsieur Roques, it really is time for you to come and see me in Laos. I guarantee you will like what I have to show you." And that is how in 2005, at the request of Francis, I came to visit for the first time the country where I would meet the benzoin tree tappers from the village of Phiengdi.

On the evening in question, I was brought into a large, dark room lit only by three weak light bulbs and a fire built on an earthen base on the floor. Villagers in northern Laos do their cooking on a wood fire in the main room of their homes, which are wooden constructions: floors, walls, ceilings—all of it is built from wide, polished timber planks. Smoke floated up into the room, depositing a thin layer of blackish tar on the dozens of baskets of every shape and size hanging from the roof timbers. Thirty-odd children had

*Northern Laos, Francis with his friend Poutao Some, a village elder and benzoin tapper, with his grandchildren, ready to follow his trade*

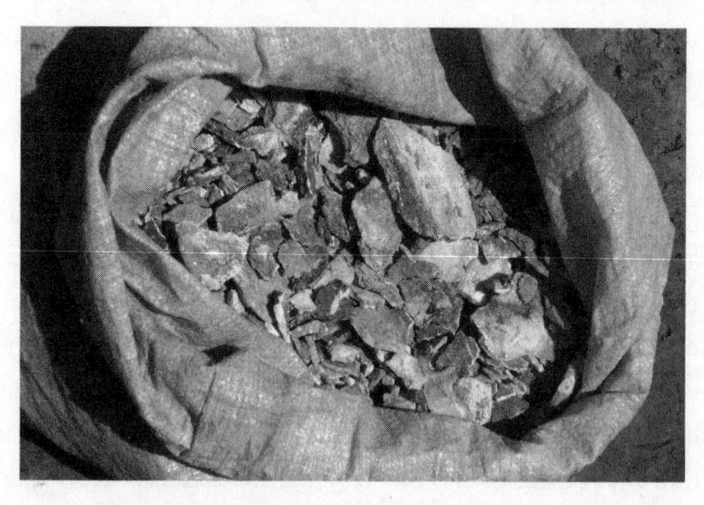

*Benzoin collected from the tree, before sorting*

settled at the back of the room, silent, thrilled to be an unexpected part of the ceremony about to take place. A significant number of villagers had gathered to welcome Francis, the man who, for the last ten years, had been buying these families' benzoin product, and the same man who would introduce me to Phiengdi, the very village with whose people he had first collaborated.

The traditional welcoming ceremony was also to celebrate my arrival at a time when hardly any Westerners were visiting the region. Francis was unaware of the fact that the village was about to offer him a gift that all of its inhabitants had been preparing for a year: a house. His house. I have forgotten nothing of his emotion that evening. The gift was certainly out of the ordinary, but from that point onward, he was one of them.

Fifteen years earlier, this French agronomist had gone to live in Laos in order to put his agroforestry theories and beliefs into action. In the country's north, he had worked tirelessly to rehabilitate and protect a significant product that was under threat of disappearing altogether: benzoin, a resin which has long been used in perfumery, and which is harvested from the *Styrax tonkinensis* tree. In order to benefit the struggling communities of the area, he implemented industry-pioneering theories and practices, fighting a relentless battle against indifference and incomprehension. He succeeded in establishing a business in an unusually insular country governed by democratic socialism. Fifteen years ahead of the times, he had recognized that our profession was going to have to acknowledge the crucial significance of the relationship—and the ethics upon which that relationship is built—with our planters, the critical and generally neglected link in the fragrance chain. He has worked in close collaboration with the benzoin tree tappers in the tropical jungles of Laos, and while the perfume industry may not realize the extent of its debt, it nonetheless owes him a great deal. Passionate, generous and charismatic, he is an exceptional man.

My own benzoin journey started in his lair in Vientiane before I had even made it as far as the villages up north. In 2005, the capital still felt more like a provincial town, with few cars to be seen on its avenues lined with tall teak trees that flanked the Mekong. Francis was set up in a sort of garage in the very heart of the city, where he lived in a small upstairs apartment. His den had a touch of the Ali Baba caves about it, filled with pieces of wood, samples of all the different raw materials he had sourced from the country's forests, benzoin samples that he had collected over the years, and an astonishing quantity of objects gathered from the Lao countryside. Baskets used for rice or fish, hunting bows and quivers, a collection of weights from the previous century that would have been used to weigh opium. His inner amateur ethnologist hoarded everything, including dozens of rice wine flasks from every era. A visit to this warehouse-cum-museum would be accompanied by a powerful smell of wax and timber, to which was added the vanilla scent of benzoin with its mild, gourmand and woody aroma. Francis's den always felt to me like something out of the 1950s, reminding me of the attics in our childhood homes where piles of objects would be stored, all with that very particular smell of the past which would be imprinted on our memories for years. The colonial architecture of the capital only reinforced the sense of time having slowed. The journey had started before we had even set off for Phiengdi.

Over the two days it took us to head up to Houaphanh province and the village where we were expected, Francis talked me through the previous three decades of Lao history, the landscape and daily life in the "Land of a Million Elephants." It is a deeply seductive country, and a peaceful one, despite the horrors of its recent history. Barely separated from Vietnam by the Annamite mountain range, Laos is as calm as its neighbor is hectic. In Vietnam, houses are built on the ground and rice is eaten with chopsticks. In Laos, houses are built on stilts and the people eat sticky rice the way you would eat bread, molding it in one hand. The two cultures are radically different. To the west, Thailand is culturally more aligned but that country is an economic giant in the region compared to Laos, whose government has stifled any growth. These days, the country's primary resources, such as timber, water and arable land, are increasingly controlled and operated by non-government entities, notably by the Chinese giant for whom Laos has become a garden annex.

We set off very early, after our ritual cup of instant coffee on the hood of Francis's pickup. The landscape rapidly became mountainous, with sharp, tree-covered peaks, gorges, rivers and, everywhere, paddy fields. Roads were full of villagers on foot, bearing crops, bamboo, chickens, items bought at market, Chinese hardware and tools for use in the fields. As we came to one village crossroads, a stone signpost resembling something out of post-war France indicated Phongsaly and China to the left, and Houaphanh and Vietnam to the right, the two main regions in the country's north that had traditionally grown benzoin. This village was a significant market town and a stopover for truck drivers. Francis stopped off there regularly and was fond of the spicy soup and small barbecued birds, migrating game birds that were trapped in large nets. He knew every plant at the market, every product, and insisted I taste everything. He spent a long time at the meat stalls sizing up the heads, feet, offal and blood before finally settling on ten or so kilograms of beef. This would be our gift to the villagers when we arrived the next day, enough meat for the *basi*, the welcome ceremony that would greet us on our arrival. At the stalls selling tools, Francis looked at every cutting implement on offer, assessing handle quality and the edges of the blades. He put a few questions to one of the vendors and turned to me, a billhook in hand. "This blade has been forged from steel from a bomb." Seeing my astonishment, he proceeded to tell me the story.

During the ten years of the so-called Vietnam War, Laos was bombarded by eight million tons of bombs, four times the tonnage of bombs dropped in the Second World War, making it the most bombarded country in history. Throughout the war, in a secret, unnamed operation, the Americans subjected Laos to continuous bombardment, targeting the renowned Ho Chi Minh trail, a significant part of which passed through territory on the Lao side of the mountain range for the purposes of ensuring supply to the Viet Cong in the south.

The Lao people were, for the most part, on the side of the communist Pathet Lao and supported the North Vietnamese, with the exception of the Hmong people, a minority group in the north who survived on opium trafficking and who would pay for their alliance with the Americans with the exile of three hundred thousand people in 1975.

On our trip in 2005, the further north we went, the more visible became the traces of those ten years. Many places had bombs on display, planted in the ground as you entered the village, shocking trophies, their tips aimed heavenward, gleaming in the sun. The amount of munitions recovered after the war was so significant that a cottage industry recycling the steel had developed throughout the country, supplying blades for tools.

I found myself captivated during that trip by the spellbinding beauty of northern Laos, the harmonious setting of the villages and their wooden houses surrounded by rice fields. Women in traditional indigo-dyed, embroidered skirts would offer their weavings for sale, smiling shyly, while children went out to the fields with the buffaloes. The notion of bombs and war against that backdrop seemed utterly incongruous, and yet it was only thirty years ago. We had spent the night in Xam Nua, in the middle of the north-eastern region; it had been a hot spot of communist fighting as well as one of the two large areas where benzoin had traditionally been produced. Francis was familiar with all the villages we passed through the next day on our way to Phiengdi, and we stopped often. He introduced me to his resin collectors, his company's warehouses and we were offered rice wine to drink in the locals' homes. Francis would ask the village elders to tell us about their life, and they would talk guardedly about the war and more openly about benzoin. Everybody in the last three or four generations had been a tree tapper. Francis would go into their homes and greet the women who were busy with their weaving, seated on the ground below the elevated floors of their stilt houses. He would ask them to show us their work and always bought some cloth, explaining to me that they did not earn much by selling it to merchants at the Xam Nua market. The people smiled at him; after ten years, he had become one of them.

An agronomist from the Charente by training and an ethnologist by vocation, Francis started his career working for the United Nations' Food and Agriculture Organization. He spent years working on livestock farming in eastern Africa, trying out programs and projects first in one country, then another. In 1989, after being sent on a mission to Laos to draw up an inventory of all the forest products other than softwood timber that might lend themselves to investment, he discovered benzoin and its long history.

It is an ancient material that has come to be known by its commercial name, Siam benzoin. Its name comes from the Arabic, *luban jawi*, meaning "Javanese incense," because there is another type of benzoin that is harvested in Sumatra from different trees that belong to the *Styrax* genus. Siam benzoin has long been recognized for its medicinal properties, as well as being used in steam inhalation treatments and perfumed pastes.

Members of the court of Louis XIV would coat their hands with benzoin paste and their face with a tincture made by allowing the so-called teardrops of resin to dissolve in alcohol. Sweet, warm and vanillic, benzoin has become a beloved classic, a favorite of perfumers wanting to evoke amber notes. It is found in most perfumes with vanilla base notes, including some of the most famous, such as Habit Rouge by Guerlain and Opium by Yves Saint Laurent.

Francis was fascinated by the subject. Recognizing that benzoin was in fact to be found exclusively in northern Laos, a remote and impoverished region, he drew up a report and submitted some proposals, but quickly realized that none of it would lead anywhere. So, with his wife's approval, he made a life-changing decision, resigned from the F.A.O. and started a business in Laos, a communist country that was shut off almost entirely from the rest of the world. The notion seemed pure folly. But Francis had done his homework. He had worked out that benzoin production was in rapid decline because villagers were abandoning the model of cultivating the trees, their harvesting was erratic, and their product was being bought at low prices by Chinese and Vietnamese collectors who smuggled it across the border, paying no duty.

Francis had a scientific model in mind: every aspect of the process, from the planting and maintenance of the benzoin trees to their harvest, needed to be integrated into the traditional growing cycle of upland rice, the bedrock of life in the northern forests. What he didn't realize was that it would take him twenty years of persistence and demonstration of his ideas before the authorities and agronomists would start to acknowledge the benefits of the local mountain farmers' agroforestry practices. By the time he established his company in 1992, however, he had well and truly identified the major challenges he faced. First of all, he had to win over the villagers by encouraging his cultivation model and guaranteeing the purchase of their harvest at attractive prices. He also had a constant battle with the black market, which he calls "informal trade," eyes ablaze with anger. And, of course, he still had to persuade the Lao authorities to allow a foreigner wanting to set up a privately owned business to move about freely in sensitive areas of the country.

Within the space of a few years, Francis had completely altered the benzoin supply chain. He had gone from village to village, offering his model. He had established hectares of tree nurseries, distributed thousands of saplings, and selected collectors whom he had prefinanced so they were able to distribute advances to the tree tappers. He had built collection centers in the heart of the regions, printed brochures explaining the growing cycle and practical details of cultivating styrax, the benzoin-producing tree. He had kept his word and purchased the tappers' produce even when he had no need for it, thereby adopting the role of industry regulator himself. He had promoted agroforestry across a vast area, restored the confidence of villagers who had been repeatedly exploited, and, before it became fashionable, implemented responsible rural development by investing his own resources and energy. He was one of the first to understand that producers of natural ingredients deserve to be treated with intelligence, integrity and a considerable degree of appreciation for their know-how.

After arriving in Laos, when the authorities finally permitted him to move freely around the country, he realized how difficult it was to access the remote villages in the northern parts. Bridges were collapsing, roads were being washed away by the rains. The benzoin regions in the north-west were only accessible by canoe. He often availed himself of military convoys to reach the more isolated villages. In the absence of any other means of transport, he would walk for days at a time, sleeping anywhere he could and eating anything he could lay his hands on. One morning, he missed an army helicopter that was supposed to save him three days on the road, but a few hours later he heard that it had crashed in the jungle. By sheer force of will, he started to collect and sell benzoin, knocking on the doors of an industry in which he knew nobody. And that is how we came to meet.

He is a forthright man, touchy from time to time, and his dealings with the fragrance industry were difficult at first. The companies operating in that sector were used to buying raw gums and resins such as benzoin from brokers, and in particular from the Germans, who had cornered that market for decades. Francis was shocked by the ignorance characterizing these deals and alarmed by the lack of curiosity on the part of such large companies, their reluctance to become contractually involved over the long term, and their apparent lack of interest, which was in contrast to his own commitment and passion. For my part, I am too fond of trees not to have succumbed to his pleas. I would listen to him setting out his vision and his plans, explaining it to the villagers and tree tappers, and admire his approach. He started appearing at fragrance industry conferences, briefcase in hand, looking like a Charentais farmer, which was a long way from the truth, and was deliberately provocative with purchasers who were a little too cocky, and whose ill-considered aim was to have him lower his prices. He was on the hunt for good contacts and, most importantly, somebody who would take an interest in his tree tappers and advocate on their behalf. He found me. We found each other.

---

When we arrived in Phiengdi, all the children in the village flocked around Francis's car. "Khampien was my first collector," he told me, introducing me to the village chief. "Together we created a nursery; he understood immediately what I wanted to do." It is a gem of a village, hidden away at what feels like the ends of the earth. Thirty houses squeezed along a trail and overlooking a precipitous valley where a small river flows below. The villagers have installed some tiny Vietnamese turbines whose propellers generate just enough current to illuminate a few dozen light bulbs.

The traditional *basi* welcome ceremony was being held at the home of Khampien and his wife, Mê Ibai, it being the largest in the village. Mê Ibai was the daughter of Phommi, who had founded the village, a venerable elder who had been responsible for showing Francis his first clusters of benzoin tears in 1991. It had taken two years for the Frenchman to obtain permission from the authorities to venture out in a Soviet military vehicle to the region inhabited by this Lao Phong ethnic minority, and Phommi had shown him his first benzoin *aliboufiers*, the pretty name given to the styrax tree by French botanists. Francis has told me about his hours-long walk with Phommi to reach the nearest trees laden with the solidified resin. "The trees seemed to be adorned with a rich, abundant exudate, as if they had just suffered a significant shock," he recalled.

Khampien gave the first speech at our *basi*. Then he officially announced to Francis that his house was complete and had been blessed, and that we would be sleeping in it that evening. Francis had a knot in his throat and his eyes were shiny with tears. He was being adopted by the village, a rare occurrence, and Francis was conscious of the honor. In an unsteady voice, he replied in Lao. Then the children came to tie little cotton bracelets—multi-colored lucky charms—around our wrists, bowing and reciting a few words. It was now time for the ritual of the earthenware pot. We were supposed to accept the invitations of the delighted guests to join in a series of good-natured duels that involved sucking up as much fermented rice wine as possible through a long bamboo straw that was sitting in a big earthenware pot. The alcohol was not very strong but I was keen not to disappoint, and I drank enough of it to make my head well and truly spin. We ate a copious amount: barbecued beef, eggs, soup and rice. Everybody sang, Mê Ibai launching into verse in her hearty voice, with the children's voices, now louder, now softer, chanting the history of the village and their hopes for the future. Francis translated; the final verses of the song wished him the longest possible stay in his new home. Then it was my turn. The alcohol had done its work. The first song to come to mind was "À la claire fontaine." The room fell silent, the children sat open-mouthed, probably stunned by the exoticism of this traditional French song. They were moments of such vivid intensity . . . The sweet-natured beauty of the villagers, the sense of shared experience attached to this grave but gentle-spirited ritual—never had I known anything like it. Later that night, when everybody had returned to their own home, Khampien accompanied us the short distance to the new house set on wooden stilts that faced the valley. Then he left the two of us to it. It was still unfurnished, so we slept on a mat on the floor of the new master bedroom. We had only one blanket between us and it was a very cold February night in the highlands of northern Laos. Regardless, these were unforgettable moments and I felt profoundly happy.

The ceremony had meant I could be introduced to the guardian spirits of the village, its ancestors and the forest. I had been accepted here now, and was free to participate in local activities. We ate the first meal of the day at Khampien's home, around the fire, and I learned how to mold the handful of rice I had taken from a handsome bamboo container in the shape of a small barrel, the most ubiquitous object in Laos. Mê Ibai fried some eggs and offered me some fat white grubs that had been collected from the bamboo. Francis watched me; he liked testing his guests with disconcerting food. I swallowed mouthfuls of the grubs with the rice; it was neither good nor bad, just the sort of food whose origins you'd rather not know. Then we set off down the track for the benzoin collection. Khampien took the lead, followed by two barefoot tree tappers. From above the village, the view took in a slope covered in dense jungle out of which loomed a number of tall peaks. It looked impenetrable. These mountains were holy, Francis told me, nobody would ever be allowed to cut down anything there. People only went there in order to pay their respects to various trees. Shamanism is very evident in these communities and the relationship with the forest is of fundamental importance. Trees are protectors, nourishing, sacred.

When we arrived at the edge of the forest, our guides fell silent. The only sound we could hear was the rustling of the leaves in the wind; even the birds seemed to respect the silence of the place. Our hike out to the benzoin suddenly felt like much more than simple research into a raw material. Everything that had happened since my "baptism" at the *basi* ceremony seemed to be a sort of silent and subtle initiation to which I was happy to abandon myself, a gift from Francis to help me understand his decision to dedicate his life to these villages. We walked for two hours, passing a number of young women along the way who were balancing baskets of corncobs on their heads, before we finally arrived at a small stand of benzoin trees. These trees can, in time, grow into imposing specimens, but they are at their most productive after seven years. The trees were grouped in an area that Francis referred to as an assart, namely the plot of land where the owner had planted them, having cleared a portion of the hillside by burning back the vegetation before planting it with rice. The benzoin cycle starts in November when the tapper cuts into and peels back strips of bark from the tree's trunk without detaching them, so as to create natural pouches where the resin can collect. It is left to exude for three months during which the liquid hardens as it oxidizes, forming the clusters that will turn into the sought-after "tears." Francis had promised me I could watch the harvesting, which had started a month prior to our arrival.

A tapper prepared to climb one of the trees. He cut a piece of bamboo and a liana, just as much as he needed, in order to use the piece of green wood attached to the trunk as a ladder rung which he could use to support himself as he worked. With a wide basket strapped across his body, a scraper in hand, he stopped at each pocket and carefully scraped away the mass of resin. Fresh benzoin is white, scarcely colored by oxidization. Over time, the teardrops turn yellow, then orange, almost brown. Fifteen or so notches had been prepared around the trunk at a height of about ten meters and the tapper collected everything from around the circumference of the incision, careful not to leave anything on the tree. As is the case for all forms of tree-tapping, the skill lies in drawing the sap from the best part of the tree without exhausting it or causing it to die. The tapper invited me to climb up. The speed with which he knotted the liana to secure the bamboo was impressive. I collected one pocket, my nose almost glued to the trunk of the tree, drunk on the delicious scent of the fresh benzoin, a vanillic bouquet, at once woody and powdery, with floral notes.

The two tappers collected the resin from the twenty or so trees in that clearing, enough to fill a fifteen-kilogram bag, but a modest haul, it seemed to me, for all those hours of work. Once back at the village, we spread out on Khampien's floor the material we had harvested and the smell of the fire was soon overpowered by the scent of benzoin. The pieces of resin, known as "tears," came in every shape, many still stuck to bits of bark. It would all be sorted into commercial grades, Francis explained to me. Khampien weighed our collection and added it to one of his bags. Over the two months of the collecting season he would be purchasing the tappers' entire harvest and would deliver it to Francis's warehouse in Xam Nua. Thanks to advances paid by Francis, Khampien was able to collect the harvest from several villages, pass it on and receive a commission. Francis received benzoin in Vientiane from his network that took in the entire north of the country, and dozens of women would sort it according to the grade or size of the tears, the largest drops being considered the best quality. He would then send his crates of product out to his customers, including to my company, and all we had to do was dissolve the tears in alcohol to produce resinoid, the ingredient which is then used in compositions. The supply chain could not be any shorter.

On the way back to Vientiane, Francis told me that benzoin was still being produced in the same regions as it had been during the colonial years. Back in what was then known as Tonkin, men would load the benzoin onto their backs in twenty-five-kilogram loads, then the harvest would be carried downriver on bamboo rafts through to the distinguished northern town of Luang Prabang. Benzoin from the eastern provinces would be hauled by columns of cattle teams through to Vientiane. There the product was sold on to Chinese merchants who would ship it onward to Hanoi, Saigon and Bangkok. The terminus of a journey that might already have lasted three months, Bangkok became the principal port for the shipping of benzoin.

All this was in the distant past, but Francis still had to navigate a complex path. "I'm a guest in this country. I must be beyond reproach when it comes to matters of tax and regulatory compliance. It has taken a long time to earn their trust, but all of it can be lost from one day to the next. It's something so many Europeans forget . . . They have been scrutinizing me for years, but now the administration and the ministers take me seriously and support me. The black market poses the biggest issue. I am the only person to have paid a levy to collect benzoin and to have paid all the local taxes and export duty. Everybody knows how the Chinese and Vietnamese go about business, with motorbikes doing the rounds of the villages, offering cash to buy product for just a little more than the price I'm paying from the very people I've prefinanced. The authorities know what's going on but the administration lets it happen because it's corrupt and it benefits, too . . ."

I returned a number of times to visit Francis. Each time I was introduced to his new projects and plans, warehouses built in other benzoin regions, different products. He now sells royal cinnamon said to be "from Hué," a bark previously reserved for kings which he pulled from obscurity after discovering some old cinnamon trees in remote temples. He has an endemic red ginger and beeswax sourced from wild hives in the forest, a unique, heavily scented wax with smoky, animalic notes, a real discovery. He works with small traditional agarwood distilleries buried deep in the countryside, and has developed a passion for the legendary *Aquilaria* tree, which is ubiquitous in Laos, to the point where he has bought a hundred-year-old tree and fenced around it to protect it from being felled. Francis practices his own form of shamanism.

When I last visited, in 2017, he was just emerging from two difficult years; he had not been spared the effects of the Mekong floods and other climate-related hazards, but he remained optimistic and determined in a country undergoing significant change. Tourists were flocking to a Vientiane that had already been overtaken by motorbikes. Sleepy Laos was a thing of the past. The Kunming–Bangkok high-speed railway, conceived of and implemented by the Chinese, was cutting a swathe through the forests of the north. The government was ordering the displacement of entire villages that stood in its way. In the agricultural sector, Chinese imperialism was silently expanding its reach, with forest concessions and the planting of enormous crops of rubber trees, maize and cassava. Unsurprisingly, modernity was starting to devour tradition.

Francis had been terribly distressed by the death of Khampien, struck down by illness in the prime of his life. "He was a remarkable man, intelligent, loyal. His grandfather used to collect benzoin, his father fought in the war. But he was hungry for progress and was ambitious. There were so many things he helped me to understand . . ." Francis had lost a friend, and a connection to the forest. Khampien's death also marked the end of the Laos that Francis had known when he had first arrived, the country that had not really changed since colonial times. It was the start of a new chapter. Francis's primary battle continued to be the fight against the black market; his position in Laos had opened up a channel of communication on the topic with the prime minister, and new laws were being drafted. "It will have taken me twenty years but we're finally getting there," he tells me. In Europe he has continued to be a tireless advocate for intelligent agroforestry. He has also been a passionate advocate for all the other fragrances harvested from trees in other parts of the world, such as Peru balsam, styrax and frankincense. We used to chat endlessly about these balsams and resins, cousins by virtue of their scent, their history and the issues that need to be addressed if they are to continue to occupy a place in perfumery across the planet.

One day I accompanied him to the opening of a high school that he had co-financed with a customer who found his approach compelling. He had patiently argued that the establishment of schools in isolated regions would be crucial in avoiding a premature exodus of young people to the cities; it would allow them to stay and to make a living from the forest's resources. We had been welcomed by a guard of honor formed by villagers in traditional dress, the young women with teeth painted red, lending their radiant smiles a mysterious air. Where I saw beauty and exoticism, Francis saw an ethnic group, the Lao Kho Poulanh, a language, a culture, a particular sort of village economy, young people able to be trained in tree tapping. Today the school numbers five hundred students, the track has been replaced by a road, and the village has electricity and a field hospital.

Visitors to the village are won over by what Francis has accomplished and his work is held up as an example of sustainable development. In his mind, all he has done is lend his shoulder to the task of conserving this resource, advocating enthusiastically for its careful exploitation. "I wanted to succeed for the benzoin communities themselves, but I also wanted to prove certain things to myself." He recalls my comment to him at the end of my first benzoin visit. "You told me that benzoin would have to be more expensive, that it was critical to its survival, a surprising remark coming from a buyer! I thought my customers would never accept it, but you were right. I realized then that buyers, too, were capable of considering their fellow men and the future."

Our industry needs men like Francis. Through his collaboration with forest producers, he has developed a way of managing an ingredient that has long been used in perfumery. By guaranteeing them an income, he has encouraged them to continue harvesting the product. That may be obvious these days, but he was doing this well before the idea had become fashionable and before other producers of raw materials started to fall in behind him, searching for the balance between respect for communities and their environment, and satisfying the needs of perfumers. Whenever I meet these committed men and women around the world who have chosen to live and work at the source of these scents, I always spare a thought for Francis, a pioneer in truly sustainable development, an architect of the agroforestry of fragrance in a far-flung corner of the world, and a tireless spokesman for and champion of the knowledge and skills of the people whose life is the forest.

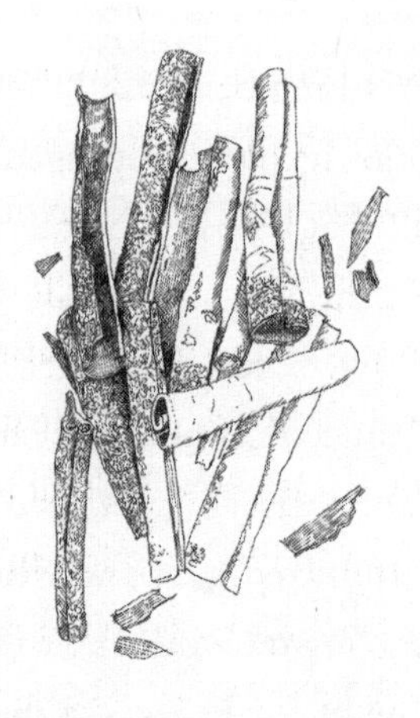

# SWEET BARK

## *Sri Lanka and its cinnamon*

"This is what a cinnamon tree looks like when you allow it to grow," Lasantha said to me, leaning against a tree of average height that looked much as you would expect a tropical species to look with its grey trunk and new, bright-green leaves. "But curiously the essential oil in its bark does not improve as the tree ages. We only work with young saplings." My supplier of cinnamon oil is showing me around the gardens of Lunuganga, on the island's south-western coast, to my mind one of the most enchanting places in Sri Lanka. I go over to the tree, peel back a small piece of bark from a branch, then rub a few leaves together, and immediately we are enveloped by a waft of that familiar cinnamon scent. The property, designed by Geoffrey Bawa, one of the country's leading architects and an instinctual environmentalist skilled in the marriage of Asian and European traditions, is an unusually harmonious space. Nature in all its flamboyant glory spills across gardens and terraces, and both the traditionally designed main dwelling of wood and glass and its annexes blend quietly into their surrounds. Modern statuary and great antique earthenware vessels punctuate my tour of the estate. Trees and flowers alike, the vegetation of Sri Lanka unfurls exuberantly all the way to the edge of the great lake bordering the ocean. It is a carefully curated setting that emphasizes the connection between landscape and fragrance. Cinnamon trees keep company with kapoks, and thin, red-trunked palms allow glimpses of a rice field edged by gnarly old frangipani whose branches trace tropical arabesques against the blue of the sky. The peace is unbroken, a gentle breeze lifts off the ocean, birds flash between trees, bundles of squawking color. For me, Lunuganga personifies the beating heart of this country whose tropical charm—a blend of exoticism and profound tranquility—is so seductive. To any visitor, the country is a mosaic of diversity and peaceful grandeur with its mangrove beaches cooled by sea winds, the grace of its dancing Buddha statues, its tea plantations blanketing the mountains in a green tapestry striped with corridors for the pickers, its flocks of birds and the scent of cinnamon. It is April 2015, and I am stopping in Sri Lanka once again, after a number of previous visits. The island is not far at all from the jasmine fields of Tamil Nadu. "You'll see! It's like India, only much cleaner!" Raja had confided to me before my first visit. "I wouldn't shout about it, but their pepper is much better than ours . . . Happily, they don't grow jasmine!"

Sri Lanka is the self-proclaimed spice island, a historically justifiable assertion given the significance of its role in the long story of

*Bentota, Sri Lanka, working on sticks to remove the precious cinnamon bark*

aromatic herbs and spices that dates back to ancient times. The island's west coast has always grown cinnamon and pepper. In this regard, it can perhaps be seen as the twin of India's Kerala coast, with its comparable western exposure to the ocean and similar latitude. These two spice coasts were the starting point of the trade which would so preoccupy first the Romans and then all Europe through the Middle Ages. The spices first traveled overland to meet caravans bearing the myrrh and frankincense of Arabia, then by sea from the first century A.D., when Greek and Roman seafarers worked out how to take advantage of the monsoon winds, heading to India to load up their vessels before returning with their cargo. These days, a map of significant spice producers would include Madagascar, Indonesia and South East Asia. However, while there may now be a number of choices as to where to source pepper, cloves, nutmeg and cardamom, the same cannot be said of cinnamon: its true home is Sri Lanka, where it proudly claims a form of self-evident pre-eminence. Cinnamon can be found elsewhere, certainly, but everybody agrees that it is not as good. The cinnamon from Ceylon, *Cinnamomum zeylanicum*, has a fine reputation as much as a cooking ingredient as for its use as an essential oil in perfumery. Its principal rival is Chinese *Cassia*, a distant cousin, produced in far greater quantity and at a cheaper price, but with a considerably inferior flavor and fragrance.

Cinnamon has always been recognized as a valuable commodity. It is one of a small number of aromatic products that has been considered indispensable since the most ancient of times. "Cinnamon" used by the Egyptians is even referenced in biblical texts: "I have perfumed my bed with myrrh, aloes and cinnamon" (Proverbs 7:17).

A resin, a wood and a spice: the combination of these three ingredients perfectly illustrates how scents were selected and blended more than three thousand years ago. The sweet warmth of cinnamon, coupled with the potency of myrrh and the incomparable smokiness of oud, or agarwood—known in ancient times as "aloes"—plunge us straight into the world of fine fragrance.

Sri Lanka's successive conquerors and occupiers, namely the Portuguese, Dutch and British, all made considerable investments in the harvesting of and trade in spices, one of the island's essential resources, until the British covered the Ceylon highlands with tea in the nineteenth century, having failed to grow coffee successfully. There must have been significant numbers of cinnamon trees, as history makes no mention of stock shortages or exhaustion of supply. It was only much later that the most productive method of harvesting the bark was developed. Quality was gauged according to the thickness of the bark, the finest coming from young saplings no more than two or three years old. And so plantations with easier access and a more productive output replaced original methods of harvesting from the branches of trees growing in the wild.

Lasantha is much like his country. He speaks in a soft voice and his manner is gracious and reserved. It took me some time to win his trust and to persuade him to tell the other side to Sri Lanka's history, the story of the two tragedies which, over the last twenty-five years, have scarred the country: the civil war and the tsunami. The civil war lasted so long that the world ended up ignoring it. The Hindu Tamils in the north were looking for independence and a partition of the country. The war against the Buddhist Sinhalese majority lasted until 2009, claiming a toll of at least seventy thousand lives. Lasantha had already invested in a cinnamon distillery in the 1990s. Together with his business partner he was distilling its bark, the source of the noble essence destined for use in fine fragrances, with its familiar scent so evocative of American apple pie and the mulled wine of Christmas markets. He was also producing essential oil from the leaves of the cinnamon tree, which are rich in eugenol. It has a strong fragrance of cloves and, for many of us, conjures up memories of the dentist. It is less valuable but is produced in great quantity for industrial-purpose perfumery, flavorings and pharmaceuticals.

In the early 2000s, attacks on the capital, Colombo, became more frequent, killing an increasing number of civilians. Lasantha told me about his rising anxiety, until the day when he was overcome by a panic attack at the thought of his wife and children taking the bus. Like so many before him, he decided at that moment to leave with his family for Australia, the most accessible country for Sinhalese. Lasantha was not in the country of his birth on December 26, 2004 when, in the blink of an eye, the tsunami carried away thirty thousand of his fellow countrymen from the southern and western coasts of the country, and specifically from the cinnamon-growing regions. That day saw his partner lose several members of his family. On my first visit to the region around Bentota, the town closest to his processing facility, Lasantha showed me the many rows of small graves along the train line which hugs the ocean on its way down from Colombo. I had a lump in my throat as I looked at these burial markers along the seashore, listening to his stories of so many of the families who used to work in the cinnamon industry and had been carried away by the tsunami. Even if for visitors today there is little trace of the one hundred thousand who perished in that quarter of a century, the wounds for locals still run deep, and in Lasantha I could sense their rawness. Very thin and of diminutive stature, he was a man of steely self-discipline who ate very little and exercised several hours a day; he was an accomplished long-distance runner. His smile belied the energy of one who has learned to overcome numerous trials and tribulations. Unhappy in his exile, he returned to Colombo as soon as the war was over to set up a new venture focusing on bark distillation, leaving to his partner the cinnamon quill side of the market, the more familiar, marketable form of cinnamon with which we flavor our food and drinks.

The cinnamon fields are not immediately obvious, hidden a little way from the coast behind the railway lines that run from Colombo down to Galle. This peaceful, coastal picture-postcard scene with its beaches and coconut palms has in the last dozen or so years once again become a tourist destination, the tragedy of the tsunami apparently forgotten. The plantations of trees belonging to Lasantha's partner are to be found around Bentota, behind the villages, a few hundred meters away from the train. Cinnamon production is organized around a traditional, seemingly immutable, model. The owner of a plantation "recruits" several families who will be responsible for every stage of the process right through to preparing the large bundles of cinnamon quills. At daybreak, the family will divide up the work among themselves: the men will cut back the saplings and lop off their branches, the wives and adolescent children will gather up all these branches into bundles that will then be sold on to a leaf-distilling facility. The saplings are in fact the regrowth from the stumps of the original trees planted. These young trees have been left to grow for three years. They are then coppiced and the numerous offshoots of their regrowth will be selected for harvesting when they have reached two years' maturity and have grown to the thickness of a broomstick. In this way, a crop can be harvested for more than forty years from the original trees. As the years go by, the stumps become very large, but the young regrowth never reaches more than three meters in height. The work is carried out in silence, the only sounds to be heard being the blows of the machete against the branches. At the end of the morning, the harvest is brought back to the farm, where the cinnamon will be prepared.

In the shade of large awnings, four families are at work, each at their own station. They sit cross-legged on the cement floor, under the cinnamon lofts, platforms of suspended wire mesh where hundreds of fine orangey-brown quills are stored, waiting to be sold. The head of the family responsible for production is a respected artisan, recognized and prized for his skills. It takes years to become an accomplished producer and to master the various procedures. The best cinnamon "makers" are sought after, they earn a good living and train their children in the trade. So, cinnamon, too, is a family affair. Each family has its place in the plantations, its role in the organization, its place under the awning. They work side by side but do not mingle. It is a fascinating spectacle, their implements seem to be from another age. At my request, Lasantha asks one of these workers about how long they have been using these methods. He seems surprised at the question and answers, smiling, that things have always been done this way. As if there were any reason at all to interfere with or alter an established and effective method. One gains access to an extraordinary world when dealing with these aromatic trees that have been valued and exploited for so long, a world of age-old artisanal skills. In the tropics I have come across implements, baskets, ropes, ladders, stone hammers, all made with materials sourced from the forest, as has ever been the case, using perfectly developed techniques whose intricate and elaborate sophistication is often beyond us. Knowledge passed down from parent to child, silent gestures as fluid as they are ingenious, incorporating an instinctive economy of resources, the beauty of objects, bamboo, bark and plaited creepers. The artisans of the aromatic tree industry, from Laos to Salvador, passing by Somalia or Bangladesh, are survivors in a world that is in the process of disappearing. Descendants of hunter-gatherers, they inspire my admiration and are testament to the best of what humankind has been able to devise in another world, a previous world. At every one of my encounters with them, when I have had the opportunity to visit, to observe, to talk, I am forced back to this troubling question: how long can all this continue?

I spend a few hours sitting with the cinnamon manufacturers, observing their gestures and tools, prompting laughter with my own clumsy attempts and, over time, initiating conversation. I could have spent days there, absorbed in this extraordinary ritual. The women are responsible for the first stage: using a knife to scrape away the top layer of green bark, which cannot be used given its high eugenol content. I'm transfixed by a young woman whose skill and speed are staggering. I recognize her: that morning I had seen her piling together large bundles of branches on the plantation. She is intimidated by me and ends up giggling. The twig is yellowish now, but will turn red as it oxidizes. In order to make it easier to peel back the internal bark, a metal cylinder is rolled over the piece of wood from top to bottom to facilitate the bark's removal. An incision is made down the length of the branch, allowing the women to peel off a long piece of bark that will then curl up on itself as it dries. The work requires a great deal of skill, it is painstaking and difficult, calling for precise gestures and a specific tool for each part of the procedure. The incision has to be made to just the right depth and in straight lines over uneven pieces of wood. The fresh strips are placed out in the sun and as soon as they have rolled up they are slotted together, one inside the other, in order to form thin rolls, each two meters in length. An older woman supervises the arrangement of the bark strips out in the sun; she is the one who decides when the rolls are sufficiently well formed. Her face is impassive as she works quickly and unerringly. Lasantha murmurs, "She lost her husband in the tsunami. He was too close to the beach." Nobody speaks loudly, any exchanges are brief and in hushed tones. The only sound is the clear, regular sound of each tool. When the best of these long pieces have been removed, the remaining bark on the branches is scrubbed clean and the pieces sent off for distillation of their essential oil.

---

Lasantha's distillery is located in the neighboring village, at the end of a small lane, hidden behind a metal gate. There is nothing to indicate its existence. It is a modest warehouse fitted out with two small distillation workshops and built around a central courtyard filled with bags of bark. Everything here operates on a human scale. The steam boiler is fed with the wood of the stripped twigs. At the end of the procedure, the workers tip up the stills by hand, pivoting them on their axis in order to empty the spent material onto the floor. It is then loaded into a wheelbarrow while new bags of cinnamon sticks are brought over to load up the tanks and begin the process all over again. Behind the workshops dozens of metal containers are lined up in rows to catch the distilled liquid dripping from the stills. It is an astonishing sight! The essential oil is further refined at every point in the process and is then collected from the surface of each vessel. Lasantha watches on in silence, dips a finger through the meniscus of the oil, nods his head. The only sound is the murmur of running water. The sun bounces off the gleaming golden surface of the receptacles; it could be a modern art installation.

What has prompted me to pay Lasantha another visit is a matter far removed from the grand scale of the country's calamities, but for a small distiller it is its own form of tragedy. For several months, my company's quality control department has been rejecting samples of his essential oil. And yet we have been working with him for years. He is conscientious, loyal and his products are normally excellent. But cinnamon essential oil is a delicate blend of a number of determining components. For olfactory and regulatory reasons, every batch must be monitored and analyzed in order to ensure it meets precise specifications. In technical terms, there must be a good balance between cinnamaldehyde and eugenol. The former is responsible for cinnamon's characteristic taste and smell. The latter smells of cloves and must take a back seat. To complicate matters, we are now required to maintain even lower levels of a molecule that is already strictly limited by the perfume industry's regulatory standards, namely safrole. These modifications call for a change in composition of the essential oils we would like to see from Lasantha and, as things stand, he is unable to meet the new qualitative profile. After numerous telephone conversations, he has asked me to pay him a visit as a matter of urgency to try to sort out the impasse; we are his biggest client and he is afraid of losing our business. Distilleries in the tropics are often small and rudimentary affairs, there being a considerable gap between the worlds of these traditional producers and the regulatory departments of large-scale manufacturers. Neither of them is familiar with the other, nor do they see eye to eye. For my part, I value having a supplier such as Lasantha who knows how to manage cultivation of the resource, how to win the loyalty of and incentivize his growers, and how to take care in the distilling process without modifying his product by adding leaf essential oil. From the producer's point of view, if we want to carry on working with beautiful, pure essence from Sri Lanka, we simply have to relax our standards. It was up to me to explain all of this to the analytical gatekeepers in Geneva who were responsible for maintaining long-term consistency of the formulae we had created and wanted to continue to produce. They could not accept the risk that an analytical shift in composition of an essential oil might modify the scent of a fragrance.

We work on the issue for an entire morning, lining up groups of five bags in the courtyard, each bag representing the usual grades of bark to be distilled. The pieces are sorted by category, the largest of them said to be the best and thus the most expensive. The last bag contains little chips and bark featherings. I linger as I register the striking difference in scent from one bag to another. Of course, Lasantha recognizes them all immediately and confirms that the percentage of the three components differs from grade to grade. Finding the balance between the scent we are after and the chemical composition whose standards we have modified seems to him too difficult a task. We spend a long time talking it over; I explain to him that the solution lies in finding the combination of each of the five grades in his final blend. We end up agreeing on a plan for some new trials. It is clear to me that it will be necessary to increase the amount of premium-grade bark, which will lead to a price issue for him and will create an imbalance in his own stock utilization. We will of course revisit the question of cost. He relaxes, and I realize he is already thinking about how to use the inferior stock that I will not be taking. It will be used in other essential oils, cheaper ones, for less demanding customers. He certainly needed reassuring and it was important that I came to see him here, in his own courtyard, to understand his challenges and to explain to him our new requirements.

The next day, I go out to the plantation again. There is no sign of the daily harvesting operations, nothing to be seen of the silent stripping work performed by the machetes. I make my way through a profusion of branches and foliage. Such a feeling of abundance is rare, and the cinnamon whose scent I am breathing in around me seems to me a symbol of the country's aspiration for lasting peace. How can one remain unmoved by this sweet-scented tree that continues to grow back, decade after decade, offering up its bark with no hint of protest, attended every morning by families who know how to treat it with care so that it can continue to sustain their own lives?

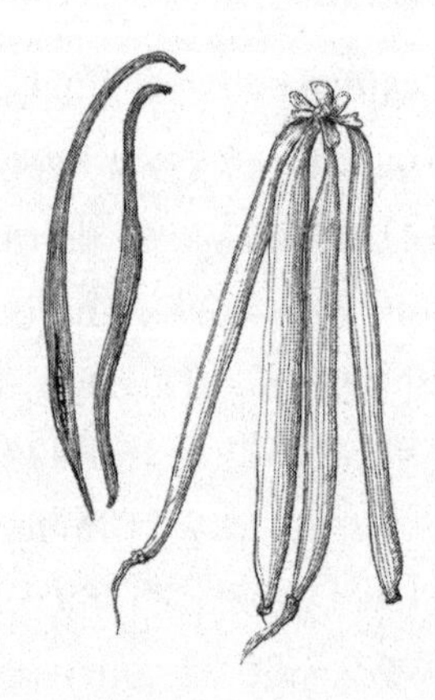

# QUEEN OF THE UNHAPPY TROPICS

## *The vanilla of Madagascar*

On the terrace of the wooden house where we have just spent the night, Pierre-Yves lights up a cigarette and lets his gaze wander across the hills that surround this village in Madagascar, hidden away in the vanilla-growing region of Sava. On the steep slopes, square paddy fields abut small plantations of vanilla protected by shade trees. “Shortages or no shortages, nothing changes for the people around here. They grow their rice to eat and vanilla to make a living, to try to make a living . . .” Pierre-Yves is from Brittany. He has the blue eyes of a sailor, is married to a Malagasy lawyer and has lived in the country for twenty-five years. I first visited Madagascar in 1994; we’ve been working together a long time. Passionate about the cultivation of aromatic plants and their distillation, he has become an expert in implementing community programs in Malagasy villages. He has been establishing cooperatives, sinking wells, and building health clinics and schools on behalf of the company that supplies me with vanilla beans. He speaks Malagasy and is very familiar with the country and its people. He is one of a small number of Westerners who know how to create useful, lasting impacts in rural areas, and as he knows all there is to know in the world of vanilla, Pierre-Yves is the man whose opinion and advice I am forever seeking out.

In 2017, vanilla had been suffering for a year from the most serious shortages in its history; the price of beans exported by Madagascar had increased tenfold in two years at most. I had come to take stock of the situation and to implement an appropriate strategy with our supplier. My company is one of the most significant buyers of vanilla beans in the industry. We turn them into extract for customers who make products with "real vanilla," of which there are very few on the market compared to those made with artificial additives based on vanillin synthetics, but which, in terms of taste, cannot be compared. We are exchanging notes on the latest news of the shortages and Pierre-Yves says, "I thought I had seen it all here in Madagascar, but this is beyond belief . . ." Ever since I've been coming to this country, I have seen much that beggars belief.

Hardly anything ever changes in Madagascar, starting with its poverty. Classified by the World Bank as among the four poorest countries in the world, it is the only one whose real G.D.P. has been declining since 1960 in a way that defies all logic. The beauty of the landscape still manages to conceal the ravages of deforestation and spoliation of marine resources. Most of the roads and

*Gigi, today the queen of vanilla, never forgot the poverty she experienced in Madagascar in her childhood and keeps giving back to the farmers*

bridges are relics of the era prior to the departure of the French in 1960. One government after another has been characterized by its inaction, and corruption in political and administrative circles is rife. Health care and education are completely inadequate, investors are put off by an absence of government regulation, and eighty percent of the population lives below the poverty line. In the thousands of villages out in the bush and away from the settled urban areas, whole communities continue to live in the same destitute circumstances, generation after generation.

Yet Madagascar is an irresistible land, and there is a seductive charm to its people, their gentle good nature, the strength of their traditions, and the extraordinary resilience that enables them to live with so little. The highlander Merina people, the country's ethnic majority, resemble their Indonesian forebears in their facial features and quiet temperament, and the Sakalavas, the ethnic group living on the northern coast, have retained their African temperament.

Even after more than twenty years, it is a memorable experience to drive through the bush villages. There are flocks of children playing everywhere with a small stone and a piece of wood, waving their hands, shouting and smiling at every passing vehicle. The women, impassive, carry water to the village from a river that may be kilometers away. It is brought back to the "huts," tiny wooden cabins with a roof of palm fronds from the voyager tree, the national emblem with its great fan-shaped leaves. Rice, the staple which features in all three meals, is cooked over an open fire, the only form of illumination after six in the evening when night falls.

There is an astonishing diversity to the landscape of this immense land. Paddy fields and endless beaches, enchanting old-growth forests, at least in the remnants that still remain . . . It is renowned for its fauna and flora: baobabs and lemurs, humpback whales that give birth off the beaches of the Île Sainte-Marie.

Madagascar, however, has her very own queen. She is not a native of the island, indeed she covered a great deal of territory before arriving from the Mexican Yucatán to prosper on the northeastern coast of the island. The queen of Madagascar is vanilla.

My primary reason for returning to Madagascar is the crisis caused by the vanilla shortage, but I also want to do a tour of the projects we are working on with our Malagasy partner and supplier. For some years now, the combination of the significance of vanilla to the fragrance industry coupled with the country's extreme poverty has forced the industry to open its eyes and invest in development initiatives. Access to water, health care, education, training . . . It is difficult to know where to start, so great is the need, but one would have to be blind or culpably indifferent to do business here without being concerned about the deprivation suffered by local farmers.

We left the facilities of Pierre-Yves's employer, our partner, the previous day. We travel first in a pickup and then, when the dirt road runs out, we switch to motorbikes, riding along paths that flank the hillsides, clinging to our young drivers transformed by the state of the trails in the bush into motocross champions. An hour out of the village we come across a surreal procession. Supervised by two men armed with rifles, a dozen porters are following a major vanilla "collector." They are making a three-day trip on foot through the bush in order to buy green vanilla pods, which will then be turned into the exportable brown beans. Every transaction takes place using cash. The collectors have to get to villages that are sometimes several days' walk away and they are carrying enough banknotes to pay for the precious beans they are off to collect.

In the Malagasy bush, goods are transported on foot. A porter carrying a thick bamboo pole across his shoulders can bear twenty-five-kilogram loads at either end. Cement, bananas, vanilla—all of it is carried on shoulders along tracks unsuitable for vehicles. But the explosion of vanilla prices over these last two years has had shocking consequences. On this particular day, the caravan of porters is not transporting vanilla beans, but rather the bundles of banknotes necessary to pay for them . . . or, to be precise, wads of notes packed into cubes and swathed in farming plastic, each packet "weighing" $50,000! Five porters, each bearing $100,000, are scattered between others carrying rice and the escorts accompanying them, whose rifles are on obvious display. The procession comprises a convoy of $500,000 in a country where a farmer's usual daily income is, at best, two or three dollars. The collector and his men are off to buy eight tons of green vanilla, which will turn into little more than one ton of black exportable beans in a harvesting season, when at least 1,500 tons must be produced to meet the market's needs. As we appear, the porters and their armed escorts have decided to stop for a break. Everything about the scene is inconceivable: the effort of these men who walk thirty-five kilometers a day bearing a load that would destroy my own shoulder in less than five minutes, the surreal value of these black parcels that line the edge of this isolated track through the trees, the obvious calm of the caravan despite their knowing that the two guards would open fire at the slightest hint of anything suspicious. We climb back onto our motorbikes behind our drivers; we are expected for dinner with the collector at the village and will spend the night at his home.

We follow the trail through a valley. Below us, the paddy fields have just been harvested. The slopes are covered in small discreet plots of new vanilla crops. The vines grow around supporting "tutor" trees that are trimmed using machetes, thereby allowing the pods to ripen in a combination of sun and shade. It is a pleasant environment that heralds, however, a forthcoming surplus. "Do you realize how much vanilla there is going to be in two years' time? Right now everybody is obsessed by these sky-high prices, nobody wants to think about it, but how are we going to manage our growers when prices collapse?" Pierre-Yves mutters to me when we arrive at the village.

We are there to open a new building equipped with six classrooms in the primary school we have financed. The project is a collaboration with the village vanilla growers' cooperative. We leave our motorcycles and head down through the village to the ceremony. We are greeted by two hundred overexcited children welcoming us in unison in French at the instruction of their teacher. "*Bonjour monsieur*, we are happy to welcome you!" The regional schools inspector was supposed to be there, but Pierre-Yves explains to me that, since receiving a motorcycle from the French foreign aid organization to assist him with his rounds, he has been too busy running his own motorcycle-taxi service to do his job. On the day the school is being opened, the government's representative is absent. I look at the children, a knot in my throat. A poignant illustration of Malagasy life. The school is brand new, there are magnificent wooden benches, but there are no books. All the children have is a notebook and a teacher, who is usually paid four to six months in arrears. Sometimes he comes, sometimes he doesn't. Today the students' parents are present. Speeches are given—the locals are fond of a speech, and we are the center of attention for the day. Pierre-Yves and I emphasize the importance of crop diversification. Our message: plant cloves or pink peppercorns to pave the way out of your reliance on vanilla crops as a source of income.

The porters arrive as night is falling and the wads of notes are set out in the collector's vanilla preparation workshop. This is where we are to sleep, on two mattresses surrounded by half a million dollars. As we share a meal of rice and eggs, the collector acknowledges that he is anxious, that nobody has ever witnessed a similar situation. Tomorrow he will divide up the cash in order to distribute it to other villages, two more days' walk away. Three porters armed with shotguns take turns standing guard and I find it hard to doze off. I'm astonished to have a little electric light from the Chinese solar panels that are now ubiquitous. Progress arrives in stages in the Malagasy bush: ten years ago saw the arrival of mobile phones, two years ago it was motorcycles, now it is solar panels.

The following day, accompanied by the sound of cheering children, we leave the school after each choosing a motorbike driver. It is raining and we have to make our way back down steep inclines to the river; it is the only possible route. The track has fallen away in several places, requiring us to push the bikes through streams of water, up to our knees in mud. Two hours later we make it to the jetty on the Bemarivo River, where canoes are waiting. These long vessels made of galvanized sheet metal, traditionally propelled by pole, are often motorized these days, with a small engine affixed to one end of the pole and a propeller at the other. It takes skill to navigate; this season, the water is no more than thirty centimeters deep in the shallows.

It is a three-hour journey downriver before we reach the next road. The rain is coming down even harder, the plastic tarpaulin designed to shelter us is no longer providing any protection. The river, punctuated by the splash of thousands of raindrops, is swollen and grey as it melts into the surrounding landscape of mountains that paint a backdrop not unlike a Chinese print. We make slow progress. Standing at the stern, dressed only in shorts, impassive and with rivulets of water streaming off him, our skipper is looking for the best channel, pole in hand. The jetty comes into view; various boats squeezed one up against the next create a large circular platform where people wander from one deck to the next, and women are cooking rice on board. We disembark into the water, soaked through, and head to the road, where we take shelter in a hut with a mother and her daughters who are selling skewers of grilled zebu beef. We sit around the coals, eating and drinking coffee, waiting for Gigi to arrive. Gigi has been my business partner in the vanilla industry for ten years. She, too, is a queen. She may not wear a crown, but when it comes to Madagascar's vanilla industry, she is royalty.

I first met her in Antalaha, fifteen years ago. She had not yet become the respected figure she is today in the Sava region, that great triangle which delineates the vanilla region on the northeastern coast. Antalaha is one of two vanilla capitals, the other being Sambava. Every exporter has their warehouse in one of these two towns. Each time I visit Gigi, memories come flooding back. After landing in Sambava, we would head down the coast to Antalaha along a potholed road which discouraged neither the Renault 4L bush taxis nor the trucks that would dive in slow motion into the enormous ruts that were carved ever deeper with each new rainfall. It would take three hours to cover those eighty kilometers. One evening a week, a section of the vanilla exporters' community would set aside any rivalry to enjoy dinner in a fine colonial house on the ocean owned by some well-known French family who had long ago come to settle on the island. Just a stone's throw from a small shipyard still building wooden dhows lived a few expatriates and members of Malagasy families of Chinese descent, all of whom were very active in the vanilla industry.

Since the departure of the French on independence, the trade and export of vanilla beans has remained the domain of two communities that run successful businesses on the east coast: the Indo-Pakistani Muslim Karana community, several families of which made their fortune on the island, and the Chinese, whose Cantonese immigrant forebears came to build two railway lines for the French in the early 1900s. I remember these dinners, where the conversation would open with a discussion about recent dramas in the latest harvest campaign and would generally finish with anecdotes drawn from the archives of the vanilla industry.

The source behind the blackish-brown pods we know so well is the vanilla orchid, a vine originally from Central America that bears a sort of plump, green, bean-shaped fruit that grows in clusters of fingers and can measure up to twenty centimeters. When the Spanish conquistador Hernán Cortés arrived in 1520 in Tenochtitlán, now only a ruin at the heart of Mexico City, he was offered a beverage which the Aztec Emperor Moctezuma used to drink. He was stopped in his tracks by the delicious taste of its vanilla overtones, which softened the bitter blend of cacao and chili. He procured some pods and brought them back to Spain. The Aztecs used to allow the pods to ripen on the vine, where they would turn yellow, split open, and start releasing their scent. And that is how vanilla would continue to be consumed in Europe until 1850.

In the seventeenth century, vines were brought from Mexico and planted in the West Indies and in French Guiana, where they were grown in greenhouses. They appeared to thrive, and flowered, but something mysterious was happening: they never developed any fruit. In 1820, further attempts to grow vanilla were made on Réunion, where experiments continued for twenty-odd years without success. Not a single pod was produced. Almost unbelievably, the mystery was solved not by university science faculties nor by the well-known botanists of the times. In 1840, on the island of Réunion, a boy aged eleven by the name of Edmond Albius, the son of a slave, realized quite intuitively the nature of the task being performed by a very specific insect in Mexico: pollination. Working alone, he started manually bringing the male and female organs of the flower into contact using a thorn from an orange tree. Miraculously, the vines started producing fruit, leading to the birth of the vanilla industry in the West through the use of a technique which is still used today. More than three centuries had been necessary for Europeans to understand vanilla.

Ten years later, a grower developed the process of scalding the pods. Dunking the green pods into water at sixty degrees Celsius for a few minutes triggered an enzymatic reaction that led to the development of vanilla's sublime aroma. Success was immediate, resulting in strong demand: by 1858, Réunion was already producing two hundred tons of pods. By 1890, an entire vanilla bean production industry had been established. Faced almost immediately with a lack of manpower on the island, French settlers established vanilla crops in the Comoros and on the islands adjacent to Madagascar, Nosy Be and Sainte-Marie.

When Madagascar became a French colony in 1896, vanilla production quite literally exploded. From fifty tons in 1910, production grew to more than a thousand tons in 1930, outstripping global consumption. In the second half of the twentieth century, the world map of vanilla production was redrawn and, leaving aside Madagascar, several countries—Indonesia, Uganda and Tanzania, the Comoros, Mexico, Papua New Guinea and India—divided between themselves what remained of the market. Others have considered it, others still will turn their hand to it, for the taste for vanilla is universal.

---

Gigi is part Malagasy and part Chinese, a woman whom life has made fiercely independent. Her mother was abandoned by her own father and married off at fifteen to a Chinese man with whom she had thirteen children. Gigi was the youngest. Her parents ran a grocery shop north of Sambava, and in the 1950s, prior to independence, they started collecting and selling vanilla. When she was six years old, Gigi began sorting beans and learning how the vanilla market worked: the different quality, grades, the drying process. At the age of seventeen, and following in her mother's footsteps, she started collecting from the villages in the bush, two days' walk from the family home in North Antsirabe. She was successful and became a collector for exporters, ultimately deciding to become an exporter herself, with a French business partner. I met her when she had established her own company, which had become one of the two most significant companies in the vanilla export market. Hers is quite a success story.

A petite woman, Gigi's expression is as determined as her voice is soft. Her upbringing has endowed her with an unrelenting drive and grand ambitions, and it is her close relationship with the farmers that sets her apart in the industry, those small growers whom she has known forever, and whose circumstances remain central to every aspect of her business.

Gigi has taught me a great deal about vanilla. She has told me of the great poverty that exists in the bush, in particular during the lean period, when the rice crop has been exhausted and the new crops not yet harvested and many growers have to sell off the unharvested and still green vanilla crop to the village grocers in order to be able to eat. Inequity is ever-present, children are malnourished, there is nothing but river water for both drinking and washing, schools are often left with no teacher, and primary health care is more than a day's walk away.

When showing me around her facilities where several hundred female "sorters" grade the beans between September and January, Gigi tells me about the complexity of the vanilla industry in Madagascar.

Grown by no fewer than eighty thousand small farmers, vanilla plants are scattered across large areas of land. Growers farm an average of one hectare of vines, which, for the most part, constitutes their sole source of income. First they will plant shade trees which serve as "tutor" trees or supports around which the vine will loop itself and then blossom. When the flowers open in October, women will work their way through the plantation pollinating each flower by hand with the help of a little bamboo spatula or the thorn from an orange tree. The grower harvests the green pods in June and July and will sell them at markets organized in the villages. There they will be purchased by collectors, often passing between three or four sets of hands before being "prepared" and finally bought by exporters. In order to obtain the best vanilla, one must have the patience to wait for the pods to ripen before harvesting them, which is to say, nine months after fertilization of the flower. The ripe pods then embark on a standardized procedure of blanching, then drying out, before they turn brown. During the four months of careful drying, enzymes form vanillin in the pods, causing them to develop the flavor and fragrance that have made vanilla a universal attraction. The last step involves the pods being sorted according to quality and size: gourmet, red, split, not split, short, "cuts" . . . Technical vocabulary for an industry that attracts a crowd of protagonists, and calls for the mobilization of significant sums of money, all in a climate of endless rivalry and competition.

Today the island produces more than eighty percent of the world's vanilla and has only taken a few decades to establish its reputation for a quality product, known as "Bourbon" vanilla, the successful identification of a product with a country. The whole world expects vanilla to come from Madagascar, even if consumers have no idea as to its scarcity. The vanilla we taste or smell in most products is synthetic vanillin; the real beans are kept only for the best ice creams and desserts.

Gigi's house is adjacent to the factory, a new facility which she has just built in the place where she was born. The buildings are bustling, vanilla is spread out on hundreds of sheets in the sun, across large, flat expanses of gravel. On the roof, a local weather scout is ready to issue a warning if approaching clouds look likely to bring rain. And if indeed rain does take place, all hands on deck: the sheets must be folded over to protect the vanilla.

Inside, three hundred women are busy sorting the beans by grade and length. They inspect each one, "assess" them, namely, straighten them out and feel them to determine their moisture content. The drying process is critical in order to stabilize the bean so it can be stored in bundles and preserved without growing moldy.

Gigi is offering advice and issuing orders in a weary voice. She has had enough of this harvest, the third in a row that is not looking good. We make our way along the sorting stations and glance at each other; I know what she is thinking. There has been widespread theft from the vines, she knows that most of the harvest has been picked before it has had a chance to ripen, meaning that the quality will be very poor, with such low vanillin levels that customers will be put off.

Gigi knows better than anyone that the vanilla industry, both production and sales, has a turbulent history, the result of recurrent crises borne of a combination of every conceivable difficulty: corruption among the authorities, speculation on the part of buyers and intermediaries, increasingly unpredictable seasonal rains and cyclones that lash vanilla crops every second year.

The vanilla industry suffered its first major crisis in 2003. A number of poor crops meant that stocks were depleted, buyers took fright, and there were sudden shortages, causing prices to skyrocket. They increased fivefold in just a few weeks. The industry ran hot, money poured through to collectors. I can still see the streets of Sambava at the time, filled with cars, mattresses, televisions, stereos. It was a veritable orgy of consumption. Vendors slipped nails into the beans to make them heavier! It lasted only a few months. Panic-stricken, manufacturers reworked their flavor formulae to reduce as much as possible the amount of natural vanilla in their ice creams and yogurts. Synthetic vanillin, at one thirtieth of the cost, would have to do the job. The richness of natural flavoring was exchanged for the poor substitute of a mere evocation of the taste. Madagascar would pay a high price for that year of shortfall. The industry's reworking of their formulae led to an abrupt drop in demand, generating reserves the following year and resulting in outrageously low prices for the next ten years. Growers were plunged into extreme poverty, earning an average income of one to two dollars a day.

Gigi did her best to react, setting up a cooperative of three thousand farmers from forty villages she knew well, having spent time with them when learning the business of collecting. She successfully sought organic certification for the cooperative, which meant that beans could be sold at a higher price. I have a very clear memory of those years when the entire industry closed its eyes to the indecently low prices underpinning its business. Despite the appearance of customers keen to invest in projects on the ground to support the growers, worse was yet to come.

We ourselves had done our best to support Gigi by purchasing organic beans, which meant better pay for the farmers. However, low prices had discouraged the other growing countries, leaving Madagascar the sole exporter of vanilla, too poor to permit itself the luxury of stopping, condemned to continue, even at such meager prices . . . A few poor seasons saw stocks depleted, setting off a fresh round of shortages, and ten years after that first crisis, the situation was once again critical. By the time of my next visit to Gigi, shortages had persisted for two years already; these days things are still just as bad.

This crisis has brought about a tenfold increase in the price of vanilla beans, the like of which has never previously been seen. It is utter madness. The amount of money that has poured into the vanilla-growing region has seen the discombobulation of almost every aspect of life. Sambava has become a permanent bottleneck thanks to the thousands of tuk-tuks imported from India within the space of a few months. Owned by two or three rich Malagasy, they are rented out as taxis. The city has also been inundated with motorbikes. Young people have set about creating a black market in vanilla and are chewing khat, the stimulant plant used for euphoric effect whose use is spreading from Yemen and Djibouti. Shops are bursting with Chinese products and the most obvious external sign of wealth is the corrugated metal being used for the roofs and walls of traditional wooden houses, even in remote villages. The cost of food is skyrocketing, and Sambava is finding itself marginalized from the rest of the country.

Gigi tells me about the widespread violence in the bush provoked by the continuing shortages. Theft of unripened vanilla pods is widespread, and, if caught, those responsible have been lynched. She is troubled that some of her best farmers in the cooperative have lost their minds and are picking the vanilla at five months, worried that it will be stolen, instead of waiting until it is fully mature at nine months. They have buried it, or put it under plastic and are now bringing it out to sell to Chinese dealers they don't know who are just passing through, or to Indians who will export it illegally and finish curing it in India. It no longer resembles vanilla. Unlike such opportunists and black-market mafiosi, people in the industry who have spent their lives handling the product are deeply invested in its quality. And yet for three years now they have had to face the nightmarish reality of acknowledging they are offering vanilla that is, at best, mediocre.

The sums of money involved are mindboggling. Traffickers are commonplace. With millions of dollars on the money-go-round, vanilla has become a playground for the recycling of dirty money. And there is no shortage of that in Madagascar, especially money from the illegal exportation of rosewood, a local species revered by the Chinese who will do anything to get their hands on it. Pierre-Yves has no words harsh enough to describe the massacre of rosewood trees, which are supposed to be protected in the national parks. Now the target of an enormous black-market industry, every accessible tree has been cut down in blatant breach of the law and hundreds of containers of rosewood logs have been exported with impunity and the complicity of the authorities. This has been going on for years and tens of millions of dollars are now being laundered through the vanilla trade. In the bush, prices are still climbing. I pay Gigi a visit at the "pre-packaging" stage: batches of vanilla blanched and dried in rudimentary fashion which she buys so she can finish off the preparation process. She is receiving millions of dollars' worth of pre-financing from her customers, a terrible responsibility when the time comes to send it on its way in bags filled with notes.

It is daunting to have to put in my orders. Prices dictate a giddying level of pre-financing, amounting to tens of millions. The finance department of my Geneva-based company is uneasy, and rightly so. I try to avoid telling them about the money making its way through the bush. Everybody is anxious, Gigi more so than anybody. Our business depends entirely on trust but I don't know how much longer this can continue.

Gigi and Pierre-Yves know that this situation will only come to an end with a fresh drop in prices. Madagascar and five or six other countries have planted vast numbers of vines, heralding a surplus in two or three years' time. Gigi is enormously worried about a possible return to famine prices. She is all too familiar with malnutrition; I have myself accompanied her several times over the years when she has gone to distribute "snacks" to schoolchildren to ensure they have at least one proper meal a day.

I assure her that things are changing, that those who use vanilla are considerably better informed of the situation at the source than even five years ago, and that, ultimately, nobody wants to make the same mistakes. I suggest that the whole industry is hoping for a return to normal prices, prices that match the growers' needs. I try to sound convincing; she pulls a face, wanting to believing me. Gigi and Pierre-Yves, both splendid soldiers, generous but weary as they continue to battle at the frontline of the vanilla wars.

Last year I returned to Madagascar. Over the past five years, some decent initiatives are starting to see the light of day. Gigi is no longer the only one; a number of vanilla users want to invest in development programs "at the source." Pressure from consumers, while recent, is nonetheless strong, and it is starting to have some effect. How are the vanilla growers treated? Are children going to school or are they being made to help their parents in the fields? Legitimate questions in the face of a complex reality.

Under rainy skies once again, Pierre-Yves is showing me around one of his recent accomplishments at a farm belonging to the cooperative. A plantation of pink peppercorn inserted between two vanilla plots, pretty trees covered in red clusters of this mock pepper that has become an ingredient of choice in the fragrance industry. We touch, we smell, we chew on the berries, children approach and mimic us, cheerful yet shy. I ask this man from Brittany about what the future holds . . . What does he make of the situation here? "For quite some time I was optimistic. I really believed Madagascar could emerge from this. Frankly, now I'm not so sure it will," he ended up replying, reluctantly, a veil drawn over his blue eyes. We've sought shelter from the rain, he lights up a cigarette. Silence. And then, "But I'm going to keep going, for Gigi, and for the sake of those kids."

I listen to Pierre-Yves. The rain has stopped, the sun is about to set. I think back to my arrival in Madagascar twenty-five years earlier. I had had to cross a stretch of the Mozambique Channel from Nosy Be Island, taking some ancient ferry for the three-hour trip. Its deck was a rusted platform just big enough to carry a dozen zebus, one or two trucks and the throng of passengers. It made sluggish progress in weather that vacillated between sunshine and tropical downpours. Standing next to the zebus, watching the sun go down amid an avalanche of cloud and color, I was reminded of the extraordinary first pages of Claude Lévi-Strauss's *Tristes Tropiques*. From his boat, which is plying the coast of Brazil, he describes his astonishment as the tropical skies are set ablaze just before nightfall, cataracts of fleeting color, changing from one moment to the next.

Sixty years later, and a long way from Brazil, I watch the tropical sky darken over this island. It is such a complicated place and yet the country remains one of the most fascinating in the world. But invariably, on each of my visits to this queen of vanilla, the unhappy tropics of Lévi-Strauss's title return to resonate with me like a constantly renewed affirmation of an endless tragedy, undeserved and unforgivable.

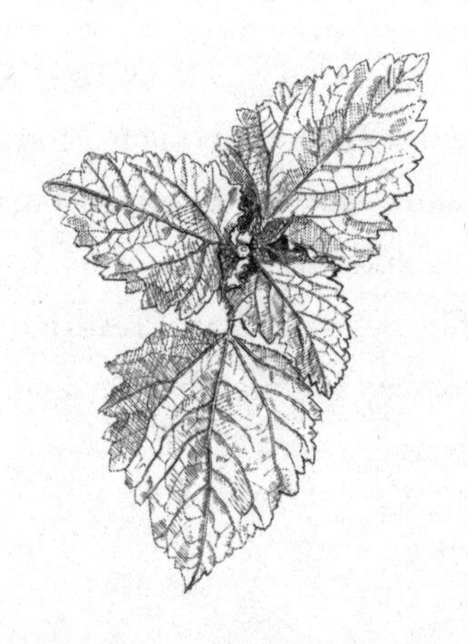

# THE BLACK-SCENTED LEAF

*Indonesian patchouli*

Athens, September 2017, the Intercontinental Hotel. Thirteen hundred fragrance industry producers, dealers and buyers of raw materials have gathered for their annual conference. The event organized by their professional association, the International Federation of Essential Oils and Aroma Trade or I.F.E.A.T., is held in a different city every year. Athens has been selected for I.F.E.A.T.'s fortieth anniversary. I am on the association's board and it is my turn to chair the conference, both an honor and a significant responsibility in this anniversary year. Delegates from the whole world hurry from one meeting to another over four days, turning the hotel into a hive of activity, where a small distiller from Bosnia or Sri Lanka might rub shoulders with buyers from the big internationals. Many of those attending have known each other for some time and are happy for the chance to catch up annually amid a tumult of cheerful greetings. At this annual get-together of the extended fragrance industry family, there is a warm, festive atmosphere that happily coexists with the theatrics of hard-nosed negotiations. In order to mark its anniversary, the association has published a souvenir book that retraces its history, the first pages telling the remarkable story of how it came to be.

Very few of the 2017 delegates are familiar with it, nor do they remember that the perfume industry's decision to set up their own professional organization was prompted by an extraordinary event, a sensational scandal in which patchouli essential oil played the lead role.

Exotic and heady, sensual and mysterious, patchouli oil has had a reputation since its appearance in Europe in the late nineteenth century for being a seductive scent with a vulgar, roguish image, according to the bourgeoisie of London and Paris. With its forceful personality, its Indian origins, and connotations of loose morals, patchouli had all the necessary attributes to make it appealing to the counter-culture of the 1970s, to the point where it became the very symbol of the era, with its bottles of essential oil, incense sticks and scented clothes. Hippies smelling of patchouli became a veritable cliché. Since first making an appearance in scent compositions at the end of the nineteenth century, the essential oil has never been out of fashion; indeed it was the focal point of a suite of great scents that were a sign of their times, trendsetters even, such as Mitsouko by Guerlain. But the launch in 1970 in a Cannes boutique of Patchouli by the brand-new company Réminiscence

*Weighing in Java: the fresh patchouli leaves have to ferment and dry before they release their amazing fragrance*

symbolized a whole new era. This unusual fragrance was so rich in patchouli that it would leave its mark.

It was a significant, iconic ingredient that made the headlines in September 1976, causing winds of panic to blow through the perfume industry. A photograph published in the *New York Times* captured a major American essential oil dealer, hands on hips, staring incredulously at a batch of opened barrels. Instead of the patchouli oil he was expecting, these drums had a thin layer of essential oil but were otherwise filled with muddy water. They formed part of a shipment of two thousand barrels from Indonesia, most of which were supposed to be filled with patchouli. The product, valued at two million dollars, had been paid for when loaded, in line with usual practice. An abrupt hike in prices at source had led to an Indonesian exporter being unable to comply with his obligations despite the letter of credit provided to him. His usual suppliers had not fulfilled their contracts, he had been left with no essential oil and had opted to engage in fraud before disappearing. It was a significant turn of events that sent shock waves through the industry. Shame and fury competed with utter astonishment.

Aghast, the duped buyers rushed to Indonesia with tempers high, issuing threats and attempting to negotiate with the local authorities. In vain. None of them managed to recover their money. The wound was deep, revealing weaknesses in the industry, and in particular an ignorance on the part of buyers as to the realities of production and collection networks. The major players did some soul searching and started looking for ways to improve control of their supply chains at source. A few months later, a handful of big British, American and French dealers met, put aside their competitive instinct which still ran deep, and agreed to form a federation whose aim was to bring together the profession and establish a code of conduct. The success of I.F.E.A.T. exceeded their expectations and it has since become a significant platform for the sharing of information; its conference is the world's natural ingredients event of the year, not to be missed.

There is universal veneration on the part of perfumers for the essential oil from the leaves of this bush grown and distilled in Indonesia. It is one of the indispensable and irreplaceable components of their palette, one of their ten desert-island ingredients. There is no question: patchouli has star status in the world of fine fragrance. But notwithstanding its prime importance to the industry, it would take more than three decades for patchouli essential oil to rid itself of its image as the problem child of raw materials. Over the years, this remarkable product has become an effective barometer of the state of our profession and a regular topic of conversation among professionals. The annual conference has always been a rumor mill of gossip surrounding the signing of patchouli contracts between Indonesian exporters and the large industry buyers. Whether discreet or, on the other hand, deliberately ostentatious, such meetings have been the subject of eavesdropping and commentary in a high-stakes environment. In particular, everybody knows that patchouli prices are fixed at the conference, prompting weeks' worth of remarks that have become industry classics: "Let's wait for the I.F.E.A.T. conference," "We'll know more after I.F.E.A.T.," or "It'll all play out at I.F.E.A.T." Skittishness on the part of industry players is explained by a climate of uncertainty around product availability and price, the result of frequent spikes in demand.

The patchouli plant is a nondescript, roundish shrub, covered in dark green, downy leaves whose scent only becomes apparent if rubbed after being allowed to ferment slightly. Its fragrance, unusually potent and utterly original, is used as much in detergents as in the most ambitious perfumes created by niche brands. Surprising and deeply seductive, patchouli makes a significant impact in fragrance formulae, its molecular complexity placing it beyond the bounds of synthetic imitation. It remains an exceptional creative weapon that is still widely used in today's compositions. Its sillage is unmistakable. Yet patchouli is also a paradox which has, over time, proved a conundrum for buyers. Indispensable and irreplaceable though it is, patchouli still seems to elude our control, with production forever shifting from one growing region to another and providing an insecure source of income to farmers while remaining patchy in quality and subject to soaring price hikes.

Having originated in India and the Philippines, it made an appearance in seventh-century China in compositions for incense. It also featured in Chinese medicine as an anti-inflammatory and antiseptic decoction, and its leaves were used to scent Chinese black ink. Indians used its dried leaves to perfume their woven cashmere products, prompting among the British a passion for the fragrance, which became a symbol of the exotic. The importation of balls of patchouli leaves allowed the British to imitate the

Indian practice with wraps and shawls, and to diffuse the scent in potpourri and mothballs. In order to satisfy the European appetite for patchouli from 1850, the British encouraged cultivation on the Malay peninsula in the Straits Settlements, which would soon have Singapore as their capital. The settlers were migrants from southern China who established plantations. It proved to be the start of the grand saga of distillation and trade in the oil which has remained the preserve of that community for a century and a half until the present day. By the 1920s, Singapore had become the center for export of patchouli leaves and distillation of its oil before patchouli took hold in Sumatra, where it prospered in the northern region of Aceh. Singapore found itself unseated by Medan, the island's northern capital, when, following its independence in 1965, Singapore embraced other ambitions. This saw Medan become the patchouli city, home to large exporters, and with trade in the product dominated by families of Chinese descent.

Back at the Athens I.F.E.A.T. conference, I encounter Petrus, a man of Chinese-Indonesian descent and one of the three major exporters of patchouli essential oil. He has been a stalwart of the business since 1967. Twenty years ago, he was responsible for introducing me to Sumatran patchouli, and for that I remain grateful. Petrus has forgotten nothing of the scandal of 1976, he witnessed it first-hand, knew all the players. Amused, he looks at the photo from the *New York Times* and murmurs, "It really was unbelievable, that whole story . . ." Petrus is the last of the big exporters based in Medan, now that production of the essential oil has largely shifted to Sulawesi. Years go by and he remains the same: thin and ramrod straight. When I compliment him, saying, "Petrus, you haven't changed in twenty years. How do you do it?" he answers with a broad smile.

"It's the patchouli. I don't have time to get old. Too much work and too many worries."

We reminisce about our first meeting all those years ago, when I discovered patchouli. In 1998, Petrus had accompanied me to southern Sumatra. Javanese families had been brought over to settle in the Bengkulu region, displaced by government policies that sought to limit Java's overpopulation. They had started growing patchouli for its leaves and essential oil. We headed into the highlands where the forest had just been cleared. Even before the arrival of palm oil, this large island was already the center of agricultural development in Indonesia. Trees were being felled and burned, land cleared, crops planted. I can still see the families of those new villages in their wooden shacks. They had seemed a little lost in their Javanese dress, silent, heads bowed. It was not of their choosing, this exile, and they were almost as foreign in Sumatra as I was. The crops, recently planted, gave the impression of having sprung up randomly, probably not intended to last more than a year. Patchouli was a passing fad, much as its growers and distillers were merely passing through. Some plots were shaded by coconut palms, others were in full sun; some were tiny, others extended over more than a hectare. The seedlings were planted in careful rows or were mixed in with other vegetables; evidently there was no overarching crop plan.

We had seen the farmers bind up the patchouli twigs into big bundles that they stacked and set to dry for a few days next to their houses in the shade of an awning in order to obtain the best yield from the distillation process. The distillery was located below the village, next to a narrow stream. The equipment consisted of three oil drums, the first used to boil water for the steam, the other two being covered with metal cones which were filled with leaves and which acted as a still. The patchouli oil was collected by the spoonful. I can still see the small Javanese women crouched in the stream, spoons in their hands, collecting the essential oil that floated on a pool of water at the end of a bamboo tube and tipping it into a plastic Coca-Cola bottle. It reminded me of the Roma's world of cistus gum in Andalusia, but I had never seen anything like it in the world of steam distillation. Petrus watched me, amused. "There must be almost ten thousand facilities like this across Indonesia. But they've only just started here, they are clueless . . . It's better on the island of Nias and on Java!"

Petrus and his colleagues would follow this migration. They would set up collection centers and offshoot sites wherever patchouli was being grown. "We receive all sorts of essential oils, which vary in color and composition. My job is to clean them up and blend the oils to ensure the consistency my clients expect," he had explained to me when we visited his warehouses in Medan. In fact, the key to the Chinese exporters' success was to collect and keep essential oil from several sources in order to create a smooth "blend." One thing that stays in my mind from my visits to Petrus is his filtration facilities. From high up in the room, the essential oil would run across a network of large, finely woven bamboo trays. I loved seeing the way moisture and impurities were eliminated from these thousands of drops of oil in this elaborate process, the end point for all those leaves harvested from thousands of Indonesian hillsides. I am still yet to experience anything approaching the intensity of the smell of the dripping patchouli oil in the tropical heat of those concrete rooms.

"I've never understood how such a little leaf can yield such a potent and complex scent," says Olivier, another master perfumer of renown engaged by our company. This star of the profession is reminiscing. "I discovered patchouli when I was eighteen as I walked past the Réminiscence boutique that had just opened on rue d'Antibes in Cannes. You could smell the fragrance wafting down the street, and its reputation for being able to mask the smell of marijuana made it transgressive and all the more appealing. Patchouli oil made up half their formula. Nobody had ever gone that far!" Olivier is the creative mind behind Thierry Mugler's Angel, which came out in 1992 and is probably the fragrance of a lifetime. It was hugely successful, a revolution in perfumery. The scent that set off the craze for gourmand notes, it is still one of the highest-selling perfumes in the world. As we're talking about patchouli, Olivier tells the story behind Angel's creation. "Vera, the brand's perfume director, was after an ultra-powerful feminine. My starting point was a personal composition, Patchou, half patchouli and some vanilla. I loved that note and was trying to work out how I was going to use it." For two years Olivier worked on transforming Patchou into Angel. He conjured up childhood memories from Alsace for Thierry Mugler, teaming praline, coffee and honey notes with vanilla, with top notes of black currant and grapefruit. "I chose not to marry the patchouli with florals;

it is precisely those gourmand notes that make it so potent." Angel's formula is ultimately a simple one, composed of twenty-six ingredients—half the number used in a typical composition—and is a quarter patchouli, a significant proportion. Olivier tells me his views on the essential oil. Patchouli has a musty scent, it is leathery, spicy, with notes of tobacco and humus, its dark, sensual facets blend with all the woody notes. It works for masculine and feminine fragrances, it has no gender, it is more than perfume, it is a drug. Back in his office, Olivier can't resist bringing out flacons of patchouli and from the very first scent strip I dip in, I am subsumed in memories of Indonesian fields. The perfumer pauses for a moment, we smell different fractions, the notes of formulae that are still works in progress. He continues, in almost hushed tones, "Undergrowth, humus, but it is also a question of color. I always conceived of Angel in blues and blacks. To me, patchouli is black. I use it if I want to inject some darkness into a perfume." This confession reminds me of the patchouli used in Chinese black ink. I love the way the perfumer's vision of the color black coalesces with the image of that scented ink leaving a sillage of patchouli on the paper.

Like the rest of his profession, Olivier has vivid memories of the last patchouli crisis in 2008. After the episode in 1976, the fragility of the production chain in Indonesia led to major shortages that also had an impact on the industry. Based as it is on tens of thousands of smallholders, the constantly migrating, cottage-industry nature of the distillation process means the product must still pass through a long chain of collectors before it reaches the exporters. When leaf prices remained depressed, farmers abandoned patchouli crops as soon as a more lucrative opportunity presented itself. This approach, combined with poor weather conditions, resulted in severe shortages in 1998 and then again in 2008. The same scenario played out ten years later with the belated recognition of the dramatic shortage of the essential oil and skyrocketing prices over a two-year period.

However, patchouli remains both a prized feature in compositions and a focal point in the industry. Buyers like to joke, "In perfumery, when patchouli supply's fine, everything's fine!" Unfortunately, in 2008, with prices multiplying tenfold within a few weeks, patchouli was not fine at all. The fragrance industry suffered a serious blow, stunned once again at finding itself at the mercy of farmers' whims, speculators' miscalculations and the Indonesian climate. It resolved to reform its strategy entirely. On the buyers' side, no more just-in-time production and delivery; all major users would ensure that they had healthy reserves, so as never again to find themselves short of patchouli essential oil.

Exporters, for their part, started prioritizing quality, stability and investment in producer communities. I recently visited some high-capacity distilleries in Java, well-designed stainless-steel constructions, examples of the new production model that buyers are keen to see implemented. Since 2010, we have been seeing farmers' cooperatives forming around these facilities, collection models established with fewer intermediaries, and contracts being signed with undertakings as to volume and price.

Together with our local partner, we have financed a pilot distillery in Java, and what I see when I visit bears no resemblance to my memories of 1998. Everything is new, clean, organized. In this deeply traditional part of the island, veiled women perform their work silently, transporting, weighing, packing. Ingenious, inventive and industrious, Indonesian farmers have adjusted to the value of their crop and these new projects mean that patchouli essential oil now offers better yields and returns than in the past. New initiatives abound. The focus remains on bringing together small producers, ensuring quality distillation and removing unnecessary intermediaries in the supply chain.

The last five years have seen spectacular changes in Indonesia. Exporters now accompany European and American buyers on visits to the growers, discussions as to distillation yield and price are no longer taboo, information is shared, and a culture of secrecy is no longer fashionable. Essential oil prices have remained relatively stable for ten years now, which, in 2008, would have been unimaginable. Patchouli continues to be a veritable barometer for natural ingredients. Transparency, responsibility, dialogue, investment at source, and respect for the growers are the new rules of the game for the perfume industry. Something resembling a combination of common sense and ethics appears to be emerging in this very fragile universe of aromatic crop cultivation, where sometimes I have the impression that the survival of certain products is hanging by a thread.

Patchouli's tumultuous past has led to various attempts over the years to establish plantations in other countries. India, Madagascar, Brazil, Colombia, Guatemala, Burundi and Rwanda have all witnessed the implementation of projects with varying degrees of ambition. But a credible alternative to Indonesia's monopoly has yet to see the light of day. It is quite rare for an aromatic plant that is easy to grow and distill not to be grown in several countries. Is this anomaly a manifestation of patchouli's own unusual personality?

With its green leaves and brownish oil, patchouli leads us into a labyrinth of black-hued stories: the darkness of the black tint selected by the Chinese when mixing the powder from its leaves into their ink, as if the calligraphers wanted its scent to accompany their thoughts onto the paper; the darkness of the note chosen by Olivier when seeking to incorporate a brushstroke of black into his perfume; the darkness of this oil's addictive appeal, tracing its potent but elusive line from ancient Chinese tradition to a contemporary perfumer's vision.

# LAND OF DARKNESS AND LIGHT

*Haitian vetiver*

Although I have always known how light in the tropics is so swiftly succeeded by darkness, I did not anticipate that my search for certain natural ingredients would bring me face to face with a number of unsettling realities. The stark juxtaposition between the beauty of the products themselves and the places where they are grown with the poverty and circumstances of the locals sometimes forces me to question the rationale behind my visits to these countries. The 2010 earthquake plunged Haiti into a nightmare. The quake and its toll of 230,000 dead added a tragedy of horrifying scale to the sorrows of an already impoverished country. I visited Haiti for the first time one year after the devastation and had driven through Port-au-Prince before taking the road west to where the vetiver is grown. I was accompanied at the time by Pierre, one of the island's major vetiver producers and our local partner, along with two of my colleagues with whom I had worked to establish a cooperative and school, which we had just launched in collaboration with Pierre's company.

To visit Haiti was like a punch in the gut. A scene of destruction, kilometers of tarpaulin-covered refugee camps, the National Palace in ruins, makeshift markets atop mountains of rubble in streets and crowds of aimlessly wandering people.

A storm blew in toward evening, causing torrents of water to gush down steep streets, creating an apocalyptic atmosphere. When we encountered people rushing through the streets as night fell, picking their way over lifeless bodies, it was as if shutters had been pulled down over Pierre's face. Death was everywhere in Port-au-Prince; both of us were stunned into silence. How could we reconcile the promised astronomic sums of international aid that had attracted lengthy commentary in the media with the ongoing tragedy of life in the capital? I felt myself faltering: the justification for my journey appeared to lose all meaning. I had come to Haiti to familiarize myself with vetiver production and to assess the impact of the pilot scheme we had launched. The brutal awakening I had just had to the state of the country rendered ludicrous any thought of trade or business, and I asked myself quite simply what I was doing there. Gripped by extreme poverty for so long, with no visible sign of progress, Haiti remains a painful enigma in the tropics. Her neighbor, the prosperous Dominican Republic, welcomes hordes of tourists, and yet on this side of the border, nothing. No tourism, no investment. On my most recent visit in 2015, five years after the earthquake, the National Palace was still in ruins.

*Digging for the precious vetiver roots in Debouchette, Haiti*

---

Within the space of two decades, Pierre has become lord and master of essential oil production on the island. He is an important figure in Haiti. He has expanded the distillery established by his father sixty years ago and it has become so successful that he regularly receives visits from government authorities and ambassadors. Now he is preparing to hand over to his children. Enigmatic, often deliberately dramatic, Pierre has a mysterious side that matches the hidden soul of his island. His passion for vetiver, much like his passion for his country, is consuming, and sometimes tempestuous. Both orator and tribune, Pierre loves to charm and he talks of Haiti with passion and conviction. He recounts the dark history of the Duvaliers, tells me about the impenetrable web of money laundering surrounding the proceeds of drug trafficking, the road blocks and hijacking of trucks carrying heating oil, the inadequacy of international aid in the wake of the earthquake, and also the bereavements he personally has suffered. He confides that he is refusing to run for the presidency of the Republic; the dream of this agronomist is to kick-start Haiti's production of lime essential oil. A thriving industry in the postwar years, lime extraction has disappeared in the wake of the deforestation that has ravaged the island and the general decline in manufacturing and production.

Pierre lives with his dogs, George, Colin and Condoleezza, in a house perched on a hillside in Port-au-Prince. There is no escaping the deliberate references in the names of his canine companions; his relationship with the United States is one that vacillates between humor and resentment, his judgments as to America's attitudes toward his country are final and irrevocable. An ambitious business man and great romantic, Pierre has a deep love for the people of Haiti, and his efforts to provide them with employment opportunities through his multiple projects are impressive. He is also very secretive, reluctant to talk of costs and productivity, and he does not mix business with pleasure in conversation. Nonetheless, I have always loved talking with him about his experiences and about his country. One evening, when I shared with him this sense of darkness and light that I felt was always present in Haiti, he replied, in all seriousness, "Never forget voodoo. It's everywhere here. It's part of our heritage, a tradition, a religion. No foreigner can ever really understand it, but Haitian life is infused with voodoo." Voodoo is a spirit cult, a set of beliefs brought by slaves from West Africa. Blending animist and Christian elements, it is practiced across all parts of Haitian society and was recognized by the state as a religion in 2003. It manifests in ceremonies where people may be seeking food relief, health assistance, or help in matters of love or vengeance . . . Celebrations involve flowers, candles, rum, sometimes bones—meaningful objects in the presence of which believers will try to enter a trance-like state. When Pierre, a man of such intensity and mystery, talks to me about voodoo, I wonder what role it plays in his own life.

Vetiver, as a word, was popularized by the success of the Vetiver perfumes created by Carven and Guerlain at the end of the 1950s. Woody and fresh, these compositions offer no clue as to the origins of the essential oil of the same name: rather curiously, it is extracted from the roots of a tropical grass. The large, ordinary-looking tufts are remarkable, particularly for the ability of their roots to bind the soil and resist erosion. Vetiver essential oil is similar to that yielded by patchouli. Both of them have really only been used since the start of the twentieth century but they have assumed a significant status in perfumery as they cannot be replicated. Vetiver's natural complexity protects it from being synthetically imitated. Produced at the south-western tip of Haiti, its essential oil is eight times more expensive than patchouli oil but only one fifteenth of the volume of vetiver is used compared to patchouli. They are thus equally significant in terms of their value to the perfume industry.

Much like patchouli, the scent of the vetiver root was already appreciated before it found its way into distilleries and enthralled perfumers. The French discovered the plant in India, its country of origin, in about 1750, and were seduced by its roots, which were woven into blinds and then sprinkled with water in order to freshen the air and perfume the rooms in the home. Working with vetiver roots is an ancient craft that continues today in Haiti and Madagascar. A plaited vetiver fan can give off a scent for months. In 1764, vetiver arrived on Île Bourbon, now known as the island of Réunion. Initial production of its essential oil dates back to 1865, with production only really taking off in the 1920s, when colonial settlers started farming it on the island, along with vanilla and geranium crops. At its peak, Réunion was responsible for a third of global production, but the Second World War would see the development of a new and distant source, which became and has remained vetiver's homeland: the island of Haiti.

Two men were the architects of this success. In 1930, a French man, Lucien Ganot, was the first to introduce seedlings from Réunion to Haiti, and within a single decade he had established four distilleries. He was succeeded by Louis Dejoie, a pioneer and visionary in the island's agricultural sector. Convinced of the opportunity vetiver could offer to local farmers, he quickly and effectively established plantations and distilleries, taking advantage of the shortages of the war years in Europe to secure supply of essential oil to the American fragrance industry. Haiti thus established itself as the world's new source of vetiver, marginalizing production from other countries.

Three years after my first visit, which had made such a shocking impression, I returned to Haiti in the spring of 2014 in order to assess the quality of some new fractions of the essential oil developed by Pierre in his distillery, and to check on the progress of our plans on the ground. The vetiver plantations are located in the west of the island, around the small towns of Les Cayes, Port-Salut and the prettily named Île-à-Vache. The hills above Port-Salut paint a stunning backdrop with their steep slopes patterned with vetiver crops, alternating stripes of leafy green and the white, chalky soil that is so well suited to the plant. A few palms and coconut trees, some tiny houses, small wooden sheds with palm-frond roofs where the roots are stored and, far below, the turquoise sea.

We headed up to the village of Débouchettes, which offers a view out to sea from the end of a dusty white track. This is the village where, four years earlier, Pierre had decided to establish a co-operative, in the hope of improving the income of local growers.

The crops have received their organic certification, farmers are duly identifiable and we buy the crop at a better price, reserving it for fine fragrance brands that are mindful of technical traceability, and of social and environmental responsibility. Part of the premium goes back to the cooperative to support projects in the village, a school being the first such project to be chosen by its members.

I had suggested to Harry that he accompany me on the trip; I needed his olfactory expertise to assess the quality of the essential oil that the facility was producing for us. Originally from Cannes, and now a master perfumer in New York, Harry was a long-time friend of Jacques and had a similarly stellar reputation in his field. Calm and curious, he was as passionate about plants and gardens as he was about fragrances. He could call forth astonishing descriptions of the smell of the logs from an oak tree he was splitting—how damp they were, whether they were rotting—and wax lyrical about the scent of fire and smoke. He could grow anything in his New Jersey garden, from patchouli to jasmine and every possible type of citrus in between. We shared a deep love of wood and trees; I was enormously fond of Harry. We had been on the island for three days and he was giving free rein to his appreciation of natural ingredients in order to conjure up for me the scent of vetiver. He was so keen for me to share the delight of his discovery of the roots. Harry likened vetiver to patchouli, two notes which, to his mind, were original, primal, profoundly human, as fundamental as fire. Both could be described as woody, certainly, but both were very different, vetiver being much warmer, and infinitely more complex. For him, ultimately, the two were a majestic combination that recreated the scent of the earth. Harry had come here to smell the roots in their soil, and there was an exultant glee to his excitement.

Together we followed the "dig" of the field: three farmers were using a pickaxe to dig up a row of clumps with long, green stalks, shaking out the root balls, cutting off the roots and separating out the stumps using a machete. Behind the men, two women were replanting those pieces that had been cut off, the beginnings of a new plant for another year's cycle, which was the time required to produce roots rich in essential oil. On the steep slopes, beneath an unforgiving sun, the work is hard, but for the farmers who have the opportunity to live off these crops, the work is a godsend. They know what they are escaping by avoiding the exodus to Port-au-Prince and its unrelenting poverty. I remember the answer given by Pierre to a journalist who was interviewing the two of us. In response to a question about the situation in Haiti, he had replied, theatrically: "Things are bad in Haiti. But in the vetiver industry things are going well. Essential oil offers a livelihood to fifty thousand families of growers."

We asked the men digging if they would lend us a pickaxe. We were keen to try it ourselves. I dug away, trying to unearth a few stumps, and quickly realized the need to eliminate any unnecessary actions. Children looked on, laughing, the sun blazed down. Harry took over from me. He looked content. He had realized a dream by joining me in Haiti, with the chance to roll the vetiver roots in his hands, to crush them, to bring them to his nose, fresh from the pickaxe, in a new garden, far from New Jersey. "It truly is the very essence of the earth. I have never smelled vetiver like that . . ." Harry was an expert in his field and a fine connoisseur of this essential oil, having created Grey Vetiver, a significant and successful fragrance for the Tom Ford brand.

In the neighboring fields, the women were following up after the dig. They would rummage in the earth with their bare hands, looking for any roots that remained after the men had passed through. Every kilogram of vetiver is precious. We made our way over to them. Seated on the ground, they were fingering the earth, shaking out any small root balls and filling a bag with them before starting the whole process again a few meters further on. I had a flashback to the *sambac* fields and the Indian worker with his wooden plow. I never tire of the hours spent with these farmers in different parts of the world. How is one to understand this industry, where the perfume comes from, without ever having picked roses in the rain or dug up vetiver from the soil? The perfumer continued smelling the roots, they would never leave his nose, he crinkled his mischievous eyes, his face glowing with happiness. We exchanged a glance, both drenched in sweat, no words necessary. We would remember this scene, the two of us, for a long time, the shared memory of our pickaxes and our hands in the grey earth imbued with the perfume of the roots.

Clusters of women in colorful dresses and shaded by parasols strolled with the natural elegance of Caribbean women along the path that ran above the field. The new school we have financed with Pierre stands alone at the end of the track. It accommodates a few hundred children from the surrounding three villages, all of them beautiful and beaming in their uniforms. It is a modest building, and ensuring it has teachers is a continuing battle. Making sure the school has a regular water supply is an equally challenging long-term project. The children love their school, and there is now a small library which we are filling with books brought from Paris. I waver between a sense of achievement and of futility when I think about this humble school and the stupendous need that results from the vacuum left by the state. It is a cruel echo of the situation on the island of Madagascar, its distressing companion in the world's poverty stakes.

Pierre's distillery is located at Les Cayes. It is the biggest in the country, the heartbeat of the town. The facility has erected a water fountain that provides water to passers-by; it is equipped to meet its employees' every medical need and also has a branch of a bank in its grounds. Vast structures house dozens of large stills that are surrounded by ever-shifting mountains of roots being unloaded by an endless procession of trucks. The roots are spread out to dry across an area the size of a football field by scores of pitchfork-wielding workers who toss enormous clumps up in the air to shake off any soil that remains from the fields. The wind blows up eddies which dance in the sun in a dusty ballet and the scent is carried this way and that in the breeze, light and warm, a milder version of the perfume that lingers in the air around the stills. The never-ending challenge for the factory is to fill up the vats with enough material to justify the cost of the steam over the twenty-four hours of distillation required for a good yield of essential oil from the tangle of brown roots. First the vats are filled with piles of dry, light roots. Then they are tamped down by a team of five men. Once the ten-meter-high stills are full, the workers clamber up on top of the material to compact it further in order to pack in as much as possible. One day I was keen to climb up to join them and we all started stamping down on the roots together, clinging on to each other's shoulders and causing widespread hilarity. The workers then started singing in Creole to accompany our stamping, laughing as they sang, a source of great amusement. I have kept a slightly blurry photo of the scene which I just love: these high-spirited workers in the still around this foreigner who cannot help but be reminded of memories of the days he spent crushing grapes in his youth.

Around the stills, everything is darker and more confined, dated machinery scattered among the facility's brand-new equipment. Here again, this shifting from darkness to light. Standing at the foot of the large stills performing their work of distilling, Harry watches the condensed water flow, rich with essential oil. At every stage of our visit to the factory, the scent of vetiver is different and he is able to find the specific, evocative words to describe it. It may smell warm, brown, woody, like tobacco, powdery earth or honey, it may have a long finish that intensifies . . . a cedary feel, dry but sweet with a touch of humus . . . Perfumers employ an astonishingly rich range of vocabulary and word associations, as if the depth of their sensory reception were compelling them toward this creativity so they might communicate and share their response.

We have reached the final stop on our visit and Pierre is opening up the room where the decanted oil is collected and filtered. This is where all the stills end up, it is the inner sanctum. The intensity of the scent literally turns my head. Harry smells a sample of fresh oil that Pierre holds out to him and he is overcome. "It is so good, it's like it has just emerged from the earth!" For these powerful, rich oils, the notion of freshness associated with the complexity of a bouquet of such deep notes is crucial, imbuing the product with something akin to an additional spiritual element.

Back in his office, Pierre offers Harry samples of different fractions, different quality, the results of a program of trials that have involved varying the distillation parameters, the age of the roots and how much they have been cleaned. The distiller is waiting for the perfumer's approval, while at the same time not providing too much technical information as to the procedures used. As ever, Haitian Pierre is keen to retain an aura of mystery.

For those few hours, I could feel Harry immersing himself in the scent of the roots, in the very actions of the farmers, in the oil which he had seen flowing at the distillery. It would all resonate with the dozens of projects he himself had on the go. He brought back a clump of roots with him in his luggage and the jar in which he keeps them has never left his office.

Harry returned to New York and I stayed on for a few days with Pierre. I spent some time observing the people in the villages and small towns, and noted the striking combination of the beauty of the people themselves and the beauty with which they surround themselves. Their clothes, the markets, the little shops, their houses—everywhere an explosion of color. Such beauty and poverty beneath the Caribbean sky . . . What is the source of the deep resilience that is so evident on this island? Haiti remains a mystery.

All of this took me back ten years, to a story that unfolded in Africa, and a happy encounter with Pierre who was a long way from home. In 2004, ten years after the genocide, I witnessed an astonishing coming together of vetiver and patchouli in Rwanda. I had come to assess plans for some patchouli plantations that had the support of the authorities. Inspired by word they were receiving about the success of projects in neighboring Burundi, the Rwandans too were keen to develop an essential oil industry in their own country. A local entrepreneur had started a nursery and had planted a few fields of patchouli, and I had been asked by one of our clients to determine the potential of these crops. One day, some advisers in the Ministry of Agriculture offered me the opportunity to observe some instruction in agricultural extension principles out in the new plantations in the countryside. When I arrived, I found to my great surprise Pierre, the vetiver man from Haiti. I had not yet visited his island but I had met him in Europe when he was presenting his oils to perfumers. Standing on top of a box at the edge of a small patchouli field, he was giving an impassioned speech to fifty or so attentive farmers on the opportunity this crop could offer to Rwanda and to Africa generally. Far from his own fields of vetiver, Pierre was a spokesman for grand ambitions. Out of nowhere he had transformed before my very eyes into a convincing orator, spreading the word. I listened to the exhortations of this Don Quixote of essential oils who was so keen to see the Rwandan farmers enjoy, thanks to patchouli, the opportunities of vetiver farmers in his own country. Firmly convinced that Africa would be the next promised land for essential oils, he wanted to persuade his audience of his beliefs and was hoping to be part of the venture.

Every now and again, when the two of us are chatting in Haiti and he raises his voice to make his point, I see an image of that Pierre once again, his powerful voice, his crusading language, standing on his wooden crate in the heart of the beautiful Great Lakes region. Whether that was his intention or not, he was, in his own way and with a hint of voodoo belief, celebrating a union between his Haitian vetiver roots and these Rwandan dreams for patchouli; he was a visionary matchmaker of natural fragrances.

# TORCHES OF THE CORDILLERA

*Peru balsam in El Salvador*

"I entered the forest, drawn in by the trees almost in spite of myself, it was irresistible. They were emitting such a powerful energy that I felt humbled, happy to receive their generosity. It was immediate, the pleasure I felt from stroking the grey bark; I saw a brown liquid oozing out of the tree, and the smell of balsam stopped me in my tracks. I stayed among the trees for almost an hour, alone, I had lost all notion of time. I know of few ingredients as addictive as Peru balsam. I love its facets, at once gourmand and woody. At first, it is vanillic and 'balsamic,' its woody notes are blond, there is nothing dark about it, on the contrary, it has a noble suppleness."

A formidable perfumer and my sometime traveling partner, Marie has just returned from El Salvador, where she came across Peru balsam hidden away in the cordillera when she was accompanying an important client. She has composed some magnificent

*High in a tree in El Salvador, hanging by a rope, burning the bark to cause the Peru balsam to flow*

*Practicing how to keep the torch burning to use on the tree*

fragrances for Guerlain, Armani and Nina Ricci, and she is one of the creators behind Yves Saint Laurent's Black Opium. She adores the woody, balsamic notes of patchouli. I like her particular sensitivity to natural ingredients, her calm, intense expression, her perceptive commentary. Sitting opposite her in Paris, I listen to her paint a picture of her balsam experiences. Sharing these intimate emotions, she leads me with her along the paths of the cordillera, evoking my own very vivid memories. I am taken by her image of blond wood. "For me, Peru balsam is gourmand, like a patisserie, intimate and warm. It is used far too sparingly in compositions; in order to do it justice, it should be used with liberal abandon!" Marie regrets the fact that allergen regulations governing fragrance components have led to significant restrictions on the permitted balsam dosage in perfumery. "But it has led me to a new project, trying to reconstitute it using other naturals, emphasizing all of its characteristics, like a picture that I would make my own by highlighting certain features in pencil." What raw materials does she intend to use? "Cinnamon, sandalwood, cedar, cacao, benzoin resin," she tells me, with a smile. "I would never have even thought of it before going into the forest."

Producers of Peru balsam are difficult to come by. They keep a low profile, deep in the mountains of El Salvador, and their facilities seem even more ancient than the great trees surrounding them. It was ten years ago, on my first visit, that I saw a balsam press for the first time, the beating heart of a production facility's equipment. The sight of machinery beneath a temporary awning, the ropes, pieces of wood, screws, beams and hoists, brought me back all of a sudden to the era of the conquering Spaniards, as if the men of Christopher Columbus or Cortés had set it up on their way through, and it had remained unaltered ever since. It lent Peru balsam a seductive air of mystery.

The modern history of this ingredient, like that of vanilla, is a tale of conquest that involves the Europeans. They discovered that the native Central Americans were using a balsam that they would collect as exudate from a tree to use as a healing remedy. The substance was effective, it smelled very good, they started using it themselves and soon it was added to the lengthy inventory of American products being brought back to Europe from the sixteenth century onward. Balsam has always been part of the local natural pharmacopoeia. It is collected from the *Myroxylon pereirae* tree, which is now found only in the mountains of El Salvador and Nicaragua. The tree has never been known to grow in Peru. It was so named by the Spaniards because it was exported from the port of Lima, capital of what was then the Kingdom of Peru. Raw materials originating in distant and mysterious places only seemed to become real in the eyes of the Europeans at their port of departure, Siamese benzoin being a case in point. Curiously, the world of perfumery has long had a tradition of adopting fanciful or approximate names. Beyond the notion of protecting confidential sources, the continuing use of these labels suggests a lack of curiosity. From the eighteenth century onward, the world of fragrances, of refinement and creativity, seemed increasingly to maintain its distance from the distant, rural worlds of its raw materials. This distance was to make the fortune of merchants and large fragrance companies, particularly those in Grasse, when they decided to establish trading posts at the source of these natural ingredients.

I returned to El Salvador in 2016 to meet up with Elisa, a young Guatemalan who had set up a company producing aromatic essential oils in her own country. Determined and talented, she had studied chemistry and perfumery in France, married a French engineer and had managed to overcome every obstacle that had been thrown at her when establishing a new business in a challenging country. With no experience, she had put in patchouli crops and a distillation facility. I was involved with setting up the business and have been following her progress ever since. These days she is successfully producing cardamom and patchouli essential oil, and over a number of years now has developed a keen interest in developing the balsams that grow in the region, namely Honduran styrax and Peru balsam from El Salvador.

Elisa wants to involve the local farmers and carry their communities along with her success; she is appalled by the poverty, illiteracy and isolation to which it seems they are deliberately confined by these Central American nations. She sources her supplies directly from producers at prices that offer a sustainable living. The daughter of a doctor, she pays health insurance premiums for workers in the fragrance industry and guarantees purchase of their entire production. As obvious as this all may seem, it remains a novel approach, and a challenging task. Elisa is determined and unstinting in her refusal to compromise when it comes to the ethics of her business practices.

I have returned to the "balsam cordillera" region, as it is known in El Salvador, to take part in the filming of a documentary for French television. It involves following an essential oil sourcing agent into a remote area and witnessing the "discovery" of a new ingredient. I was initially reluctant, doubting television's ability to recreate these stories without distorting them or misrepresenting them in order to paint an attractive picture at any price. In the end, I let myself be convinced that it would be a good opportunity to tell the story of the tappers and their extraordinary work, and I agreed.

We have been driving for six hours from the head office of Elisa's company in Antigua, the former capital of Guatemala and a colonial architectural jewel. Her cooperative is hidden away in the highlands above San Julián, reached by following a narrow track all the way up through the tropical vegetation to the producers' rudimentary buildings. The director of the cooperative is waiting for us, along with a group of tappers, respectfully lined up with hats in hand. On the horizon, through enormous clumps of bamboo, a ridge of forested mountains can just be made out: we are in the heart of the balsam cordillera. The production workshop looks out over a jungle-filled valley, out of which appear one or two *Myroxylon* at twenty-meter intervals. These imposing, balsam-producing trees are twenty to thirty meters tall, and at least eighty years old. Over time, they have grown into astounding forms with grey, striated trunks, the result of decades of balsam production.

The tappers are independent contractors engaged by the owner of the trees, with whom they share the results of the collection process. Fifty-year-old Franklin is an experienced collector whom I got to know on my first visit. With dark eyes in an emaciated face, he is as thin as a rake and never without his white hat, which accompanies him wherever he goes. He started climbing trees at the age of fifteen, continuing the work he learned from his father. In his musical Spanish, he tells of the risks and hardships of his job. Tapping for Peru balsam is perhaps the most impressive of all the different jobs I have come across in my travels. Franklin readies his equipment on the ground: ropes, a swinging seat, a cardboard fan, a knife, a packet of rag cloths and a bundle of sticks that he will use as a flaming torch. The sticks are taken from the *Myroxylon* tree, a slow-burning wood that produces excellent embers. He lights the bundle, waits until it is glowing, loads it onto his shoulder along with the rest of the equipment and heads over to the foot of a tree, a plume of smoke fanning out from his neck. A single throw of a rope into a high fork allows him to start his barefoot climb. Fifteen or so meters up, he sits down on his little seat, suspended over the void. The torch is still smoking behind his shoulder. Now he can set to work. In order to obtain the balsam, he has to stimulate the tree. Franklin slices out a piece of bark and peels it back from the trunk, then takes the torch and uses the fan to kindle some flames. Still seated, feet pushed up against the trunk, he holds the flaming torch to the exposed wood and the surrounding bark, moving it backward and forward, singeing it to encourage the balsam to flow. These rituals must be witnessed first-hand in order to appreciate the true nature of the task of these fragrance "hunters" and their life in the forest. Now Franklin applies rags to the burnt area, affixing them to the edges of the peeled bark. In two or three weeks' time, they will be saturated with balsam and he will climb back up to collect his harvest, some of which will have gathered in the rags, the rest in the pieces of bark. He goes through the same process again in a dozen or so places judiciously spaced out along the trunk. The Salvadoran tapper, much like his Lao counterpart, knows exactly how much he can ask of the tree without endangering it. Any tree that is tapped in one twelve-month period will not be tapped the following year, received wisdom that allows hundred-year-old trees to continue to be exploited. The collectors know how to manage their resources. Their lives depend on it.

The harvest is brought back to the workshop, a cement slab with a tin roof. Here the precious scented liquid will first be extracted from the rags and strips of bark by bringing them to a boil, pressing them and subjecting the liquid to a process of purification. Then Franklin will collect the results of his tapping. The press is operated manually: a man pushes an enormous piece of timber that serves as a lever, which in turn squeezes a basket made of rope and cable out of which starts to drip a liquid blend of balsam and water into a basin. The procedure is carried out first with the rags and next with the strips of bark. The blended juices are then heated up until any water has evaporated and what remains is a pure syrup, whose viscosity is measured by pouring a few drops onto a piece of glass. The raw balsam has a delicious vanilla perfume, warm with caramel notes. Once transformed into an oil or resinoid, the persistence of its scent makes it a remarkable fixative in a composition, easily married with floral notes or sandalwood, with which it makes a particularly fine accord.

The ceremony of the pressing process is as impressive as the work in the trees, a spectacle that remains untouched by the passage of time. It is as if, over the ages, a balance has been struck between method of production and yield in order to extract only the most desirable component of this natural ingredient, a balance that nothing must disturb. How is it possible that there are still people working in such conditions?

Having climbed back down from the tree, Franklin is smoking a cigarette in the shade and telling me about the nature of the work. The danger, first of all. Falls are rare but do happen, mainly if the hook holding the plank on which the tapper sits should snap. Elisa has spent two years replacing the old pieces of the concrete reinforcing bar with new steel hooks and her company has taken out health insurance for all workers belonging to the cooperative. For the young boys in the area, climbing trees is almost their only possible option. "If you want to be able to eat, you have to make balsam," Franklin says. "So we teach the young men how to tap the trees, but they have to know that their product will be bought, and they have to be sure of the sale price in order for them to really commit to it. There have been so many years of rock bottom prices . . ." Over time, product quality on the market has declined and prices have continued their uninterrupted fall. The industry has squeezed the tappers as if they were their own balsam rags. In a display of disregard and negligence, European and American buyers have left the industry in the hands of brokers and local intermediaries who do not care about the fate of the cordillera laborers. This is where the story almost stopped. Severe balsam shortages resulted from the drop in production levels, and this in turn led to two years of steep price increases that alarmed buyers. Elisa was ready, with her cooperative, her purchase commitments and clear communications with her customers: a pure product, its origin guaranteed, was possible, provided they were prepared to pay the price. In her mind, it was not possible to separate product quality from the remarkable work of the tappers. There was no question of guaranteeing one without overhauling the other.

It took three days to complete filming. The team was keen to focus on the heart of the story: the sourcing agent who had come to find a new product, unsure if he would succeed in his quest . . . I was feeling as though I had been pushed into the "Tintin in El Salvador" role that I had dreaded from the outset, until Elisa and I had an idea: bark balsam.

Strips of bark still have a very good scent even after they have been put through the press. It results in an astonishing reconfiguration of the fragrance, adding a floral note to the classic character of the balsam. It would be possible to do a trial alcohol extraction from the "spent" strips of bark using the equipment that had just been installed in Elisa's facility, and we would be the first to do it. This approach made sense, and I liked the idea of bringing back a new sample to Marie, a little piece of this tropical forest of which she always spoke so fondly. It was moving to hear Franklin and the young apprentice tappers speaking to the camera. The director was captivated by the beauty of the shots among the trees and around the press. After a few sequences recorded in the Guatemalan facility to capture the trial extraction, the team was able to film the departure of Tintin the sourcing agent, sample in pocket.

For the final scenes filmed back at our premises in Paris, I showed the sample of "bark balsam" to Marie and she loved it; it was different from classic Peru, with a new facet that added a floral touch blended with woody warmth. I watched her smell it; it was evident from the look on her face that it was bringing back memories of her experiences in the forest. Perfumers often prefer a novel derivative from a familiar ingredient, a new note they can introduce into the familiar architecture of a composition. When we talk about her work, Marie always emphasizes the difference between smelling a flacon and sniffing an ingredient after touching it, picking it, scratching it at source, where it is growing. There is something similar about all of the perfumers' reactions when they are in the field. Fabrice, Jacques and Harry . . . they all envy this part of my job and are happy, in that moment, to take on the job of sourcing agent themselves.

I returned to Central America a few months later to visit Elisa and her husband, Jean-Marie. Their business is growing and developing. Jean-Marie is always on the move, scouting for styrax in Honduras, for pink peppercorns in Peru. They have discovered an ylang-ylang plantation in the Guatemalan jungle, they are passionate about Mayan vanilla and they dream of finding real Tolu balsam, buried deep in Colombia.

Their enthusiasm is contagious. Their intelligence, their energy and their ambition are defining characteristics of the naturals producers of the future. They are convinced that treating and paying the locals well is the key to their own success, and as I listen to them, I think back to Francis in Vientiane, of course. Here, on the other side of the world from Laos, they are walking in his footsteps.

Elisa had warned me that the situation in El Salvador was changing. The feared Salvadoran gangs, or *maras*, who normally confined their activity to the drug trade and crime in the cities, appeared to be showing increased interest in the balsam trade. They had come up from San Julián to threaten and extort traders, and a month before my arrival, the inevitable had happened, with the first gunshot fatality. With admirable calm, Elisa told me about the *maras*, describing the situation like a temperature spike that would disappear in the same way it had first appeared. El Salvador is a violent and wretched country, the most repressive in the world when it comes to abortion rights, for example, and the lot of women there is particularly distressing.

We retraced the long route back into the mountains and up to the production unit where Franklin was waiting for us, his sons beside him. We gathered around the press to drink and chat, the basins steaming, balsam heating up. I thought back to Marie's accounts and suddenly felt like loading onto my shoulders a bundle of sticks that was to be used as a torch the next day, and setting off to see the trees. Franklin followed me; he found it amusing to see me carrying an unlit torch. We stopped at the foot of a fine-looking specimen. I told him how much I admired his ability to carry burning wood on his shoulder. "You know, the cordillera torches are our life. Lighting them carries with it the promise of balsam, it means earning enough to buy food. Unlit they serve no purpose. At least when they are lit, they allow us to earn a living, provided we don't get burnt."

# THE SACRIFICIAL FOREST

*The rosewood of French Guiana*

There were an estimated twenty million elephants in Africa prior to the Europeans' arrival. Inventories carried out in 2014 found that only two percent of those magnificent creatures remained. At the current rate of slaughter—twenty thousand animals a year—only one percent of the elephant population that existed in Africa when the Europeans arrived on the continent will survive. Within two short centuries, humans will have succeeded in eradicating ninety-nine percent of a species generally considered to be one of the wonders of the animal world.

In the plant kingdom, a single century will have sufficed for the Americans to have achieved the same result with *Sequoia sempervirens*, the magnificent giant redwood of the north-western United States. These sublime trees, nature's greatest creation, capable of living more than two thousand years, probably constitute

*A rosewood distillery in Cayenne in 1910*

*(Review, "La Parfumerie Moderne," 1920)*

the finest old-growth forest on the planet. The need for timber as a building material in the Californian gold rush of 1849 gave the green light for the felling of a mountain range as large as Corsica, which continued through to the 1950s, leaving only one percent of the original tree stock on the ground.

The elephant and the sequoia, icons of beauty in the world, are also dreadful symbols of the relationship in the nineteenth century between explorers and colonialists and the natural world. It was a world that had only just been mapped. Nature, so resolutely dangerous and hostile, was considered a limitless resource. It became the object of large-scale conquests, attracting insatiable appetites. Eyewitness accounts abound as to the reaction of hunters in Africa and America when they saw the enormity of the herds of animals of every species. Not even the least ill-intentioned among them had any notion that a species—elephants and bison, for instance—might be finite. We would have to wait another one hundred years before the photographs of the tally boards that recorded the massacres would start alarm bells ringing. It was the same story for the American lumberjacks, whose story I find particularly moving on account of my father's brief involvement in that world. In 1950, he was felling trees in Klamath Falls in California. Even then, some of the trees logged were old-growth sequoias. This man, who was so passionate about trees, told me how every ounce of energy had been focused on dealing with the harshness of the conditions, felling and skidding these monstrously large logs, and that the idea that the forest might one day be logged out never occurred to him or to his workmates. This symbolic one percent resonates with me as if I were standing on the edge of a precipice. It is one final alarm bell. The surviving redwoods are now protected in breathtaking national parks. Will the vertigo induced by that one percent be enough to save the elephants? Is there a threshold beyond which, whether consciously or not, man will lay down his weapons of rapacity, illegal trafficking, poverty and unconscionability?

In its own context, perfumery has its share of elephants and redwoods, its stories of mismanaged and depleted resources. I arrived in Cayenne in 2002, close on the heels of one such story, perhaps the best known. The destruction of French Guiana's rosewoods in the first half of the twentieth century had largely faded from memory when in 1997 a minor news item hit the media, putting the species back in the spotlight. An N.G.O. going by the name of Robin des Bois launched a press campaign accusing a prestigious luxury brand of contributing to the deforestation of the Amazon by virtue of the rosewood oil present in its most famous fragrance, No. 5, a fragrance known the whole world over. The finger had been pointed at the brand for its use of oil from a tree at risk of being wiped out. An article appeared on the front page of the daily paper *Libération*; a story about the luxury industry destroying nature would be certain to sell copies, but it was also indicative of the public's growing awareness of broader environmental issues, where the Amazon and its deforestation have always featured prominently. The amount of oil and trees involved was, in fact, very small, the equivalent of four to five trees every year, but the impact of the media coverage on the industry's image was quite devastating. The fragrance house took the matter very seriously; the two parties entered into talks and finally reached an agreement in which the company using the oil undertook to plant a significantly greater number of trees in the Amazon than they were exploiting. My company was charged with implementing this undertaking and we in turn tasked a branch of French Guiana's National Forestry Office with the job of planting four hectares of rosewood.

Four years after implementation of the project, I had come to Cayenne to inspect progress. The idea of replacing aromatic trees was still a novel one, but it was my feeling that replacing wild resources that had been exploited with new stock would inevitably become a major issue. Increasing awareness of the fragility of tropical forests would allow the perfume industry to become the ideal testing ground by virtue of the moderate scale of its requirements. Thus, this modest project in Cayenne represented a first step in reversing history, one actor in the perfume industry symbolically replanting the tree that the profession had had a hand in wiping out sixty years earlier.

The name "rosewood" applies to several species of tropical trees, referring to the color of their wood or indeed to their scent. Rosewood from Madagascar has no fragrance, but this has not prevented it being overexploited, given the desirability of its qualities in cabinet-making, which the Chinese adore. Europeans were aware of another South American species known as rosewood, which, from the seventeenth century, was sought after for the beauty of its wood, used to great advantage in marquetry work.

I am referring here to the *Aniba roseadora* species, whose female tree has a unique characteristic: its wood is rich in an essential oil consisting of more than ninety percent linalool, a fragrant component that is very common in natural products, in particular in lavender and bergamot. The oil from this wood has a divine scent, a characteristic freshness in its natural version that is elegant, subtle, sweet and far superior to synthetic linalool.

The wood thus made its way from the hands of woodworkers to perfumers' noses. Its oil was first distilled in Grasse in 1875 and earned prominence in fresh bouquets as a top note in fragrances. Its success sparked a hunt for rosewood in the colonies. The first logs arrived from French Guiana: it has always grown well in this northern part of the Amazon, where one finds the trees that are richest in the best-quality oil. It was not long before two French companies had set up in Cayenne in order to exploit this wood, loading it onto schooners, then unloading the first hundreds of tons of stripped tree trunks onto the docks of Cannes.

The emergence of this oil—assumed to be freely available—proved a windfall for the burgeoning European fragrance industry, and it started to be widely used in compositions. It was so successful that attempts were rapidly made to streamline trade in the product. The wood was first distilled in Cayenne in 1890, in stills normally used to produce *tafia*, the local alcohol distilled from sugar cane. Production remained at moderate levels until 1900, with one or two tons of essential oil produced every year, but then production rates took off. In 1912, seven distilleries swallowed up five thousand tons of rosewood to produce fifty tons of oil. Extraordinary methods were gradually implemented to source these thousands of tons of wood from the forests, methods that ultimately would have a devastating outcome.

In the middle of the nineteenth century, prospectors and loggers started working their way into the forests of French Guiana. Their only access to the interior was by river, the Approuague and Oyapock rivers in particular. They were mostly interested in looking for gold but results were unreliable and these forest explorers soon turned their attention to balata, from the *Manilkara bidentata* tree, a less elastic sort of rubber than that produced by the *Hevea brasiliensis* plant but with insulating properties that were of interest to the rapidly expanding electricity sector. Accounts from the early years of the century reveal clearly enough that the practices used were a far cry from any form of respectful "tapping": trees were bled dry in a single go, then left to die on the spot. The forest was considered akin to a mine, a sentiment reflected in the way it was treated. Demand for rosewood would see a vast increase in the systematic exploitation of the most accessible tree populations, namely those closest to the rivers. These great trees had to be felled and chopped into logs weighing fifteen or so kilograms each, a not inconsiderable task. Once there were sufficient logs piled onto the banks, the loggers would create a dam across the river, toss in their harvest, then release the reservoir so the timber would be carried as far as possible downstream by the force of the new current. Log driving was a well-established forestry technique, but in the conditions presented by the French Guiana forest, it must have been an extremely dangerous one. It was chain-gang work, both figuratively and, sometimes, literally, for the rosewood loggers who had to deal with fevers, snakes and scant food supplies. Escaped convicts from Cayenne seeking refuge in the forest, and volunteer convicts who had put their hand up to work, these prisoners also played their part in this story. For several years, the exploitation of lumber closest to the coast and most accessible to the banks of the rivers was productive and cost-effective. But the pace was such that the loggers had to advance deeper and deeper into the forest. Trees were now being felled at a distance of three or four kilometers from the riverbank. To the initial labor was now added the need for the men to transport the logs on their backs, four or five logs at a time, stopping every thirty meters. It was debilitating work.

After the Great War, an increase in demand for rosewood oil saw the patenting of a new invention in sourcing supply: the "floating hydraulic distillery." A barge capable of crushing the logs and distilling them on the rivers in the heart of the forest meant that only the oil itself—and not the wood—required transportation, a vast saving. The machinery arrived from the city and by 1926 there were up to ten floating distilleries. It was a record-breaking year with a production of more than one hundred tons of oil. I try to picture for myself the felling, the grinding, the distillation of ten thousand tons of wood in the Guianan forests. The harshness of such working conditions, of that life, is to a great extent impossible to conceive these days. It is difficult to project oneself into the world of rubber tappers, of rosewood loggers, men called upon every day to pay an often inordinate toll when faced with the vast scale of the Amazon forest. These "miners" of the forest were its first victims, merely executing the plans of an industry that had chosen not to concern itself with the management of its resources.

In the 1920s, French Guiana was not yet aware of the fact that the rosewood oil motherlode was starting to run dry. By the end of the decade, production dropped sharply; the trees were increasingly inaccessible, too rare and too expensive. The industry survived through to the Second World War, but then production halted almost entirely and Cayenne's last distillery closed down in 1970. Distillation was officially prohibited in 2001. It had taken fifty years for the resource to be depleted. The increased availability of synthetic linalool would compensate for the lack of rosewood oil, a reflection of the significant shake-up of the fine fragrance industry in the 1970s caused by the widespread replacement of natural ingredients with synthetic molecules.

After the war, Brazil did its best to take over production with an oil produced from a species related to the Guianese tree but of inferior quality. That country, too, would soon see its rosewood populations similarly depleted. Increasing awareness on the part of the authorities led to attempts to mandate replantation, previous mandates having been ignored, and this was followed by the imposition of dramatic restrictions on felling when the tree was listed by C.I.T.E.S.,* the body responsible for regulating trade in endangered flora and fauna. Production in Brazil is now tightly controlled and has diminished to negligible quantities. And so,

* Convention on International Trade in Endangered Species of Wild Fauna and Flora.

much to their great regret, rosewood oil has disappeared from the palettes of perfumers.

There was no evidence of any of this when I arrived in Cayenne. Leaving the city, I had the feeling of being in some Amazonian suburb. All trace of old-growth forest had long since vanished from the landscape, with only sections of secondary forest appearing here and there, some parts taller and denser than other parts. A forestry technician was to be my guide for a visit to a plantation an hour out of Cayenne. Rain was bucketing down when we arrived at the demonstration plot, and the forester used his machete to cut a path toward the trees through the sea of vegetation that was rampant on account of the rainy season. Tips of new saplings poked out in ordered lines through the bushes and vines. They might have been four meters tall, their trunks no thicker than the handle of a pickaxe. Not every tree grows quickly in the tropics.

Rosewood, fine grained and dense, grows slowly and should never be felled before it reaches the age of thirty. I pulled back the weeds and vines as I walked, soaked through. Standing by one of the young saplings, my thoughts turned to the confronting tally of those thousands of tons of trees it had taken barely more than a century to cut down so savagely.

The Amazon's wealth of biodiversity feeds many fantasies in industries that rely heavily on plant-based materials, be it food, cosmetics or pharmaceuticals. The world of fine fragrances is no different, with perfumers regularly asking me which new products from wood, flowers, berries or fruits might emerge from the forest to be added to their palette of ingredients. It may come as a surprise, but despite all the research and trials, the Amazonian raw materials most used in fragrances remain limited to three oils sourced from the forest. The tonka bean tree is still used because of the value of its fruit. *Copaiba*, or copahu balm, is harvested without having to fell the trees, by periodically boring into the trunks to make the fragrant sap run, which does not endanger the tree. Rosewood scent is stored in its fibers, the source of its own misfortune, for it means it must be felled in order to capture it, which seals its fate.

It will have taken decades to move from a mindset of "forest-as-mine" to "forest-as-garden." What was to become of that new plantation? Was it just a knee-jerk reaction to attacks by the media or indeed the beginning of a commitment to the possible renewal of distillation in French Guiana? There was no questioning the brand's desire to do the right thing. It had been investing in rose and jasmine crops in Grasse for years and was pioneering the revival of the perfume industry's historical traditions. As for the rosewood trees, it had been a decent response to prioritize the planting of more trees than they were responsible for removing, and to monitor their growth. But the brand's own essential oil needs were too limited; they were being met by supply from Brazil, which meant that no new distillery would be established in Cayenne. Ten years after my visit, other rosewood plantations started to appear in Brazil. Several initiatives tried to promote the distillation of branches in order to avoid cutting down whole trees; better to opt for pruning rather than felling.

But while waiting uncertainly for a plentiful amount of good-quality linalool to be produced from branches alone, most perfumers are sourcing a lesser-quality linalool from other natural ingredients, or contenting themselves with the monochromatic, flatter scent of the synthetic version. Many of them regret the disappearance of the unrivaled rose-hued elegance of the Guianan oil; its reappearance in Brazil in a form that could be sustainably exploited would be welcome news.

Nowadays the perfume industry is focused on investment in growing and preserving natural ingredients, but efforts remain feeble when it comes to the forests. The desire to plant aromatic trees quickly runs up against issues of time and thus profitability: how long must one wait before they can be exploited? Will we see a rebirth of the original version of rosewood oil after all? Who, these days, will risk large-scale planting of standing linalool, which will not be harvested for twenty or thirty years?

Twenty-five years ago, my father set about creating an arboretum, a collection of trees that he planted over a few hectares in the Landes region of France. One of the first trees he wished to plant, right in the middle of his property, was a *Sequoia sempervirens*. Far from its native land, the redwood accepted the challenge, and grew, and today it stands twenty meters tall. My father watched it grow with his head full of memories of his work as a logger in California, and thus he paid it particular attention. It was a symbolic replanting of a tree that, at another time in his life, he had cut down. He was always asking me for the latest news about my proposed book, so I told him what I was hoping to write about aromatic trees. "Don't forget to say that all forests grow back, either on their own, or with some help, trees don't bear grudges, they simply have more time than we do." As difficult and slow as the process may be, I would like to believe in a possible reconciliation of perfume's ephemeral nature with the longer lifetime of the world's great trees. And I would like to believe in the survival of the elephants.

# IMPENETRABLE RIVERS

*Venezuelan tonka beans*

Our dugout canoe was heading back up the Caura, a tributary of the Orinoco in Venezuela. In the humidity-saturated heat, the grey tones of sky and water mingled in the raw light. White egrets took off ahead of us with shrill cries. I watched as the riverbanks slipped past, lined with trees of every shape, their enormous roots plunging into the river, and every peculiarity of the landscape reinforced the sensation I had of having penetrated the very heart of the Amazon. Every now and again, an enormous tree would loom out at the river's edge, its trunk thirty meters tall, its giant crown clearly outlined against the sky. We sailed past towering foliage, from which appeared branches covered in orange and yellow flowers. I thought of everything I had read or heard about the planet's great green basin. A commonly held fantasy of a jungle paradise, trees rippling into a distant horizon, a wealth of biodiversity yet to discover. But also the crude reality of unrestrained deforestation leading in turn to the disappearance of indigenous peoples whose gravest error was to have lived there for as long as anybody could remember. It was an issue that had preoccupied me from the moment I started working in this industry: how should we manage the exploitation of these aromatic trees, both the trees we fell and those we conserve? As we made our way down this Amazonian river, the issue came sharply into focus.

In the vast catalogue of natural, fragrant raw materials, the tonka bean is an astonishing example. The fruit of a tree that grows wild through the forest, it finds itself subject to an assortment of handicaps and risks that ought to have prevented it ever forming part of a perfumer's palette. Depending on local conditions and the vagaries of the climate, the tree may unexpectedly decide not to flower, sometimes for several years in a row. The blossoms must then bear fruit, though again there is no guarantee this will happen. If the trees do indeed bear fruit, collection of the beans is dependent on communities for whom the beans are only a supplementary source of income, at unpredictable rates. For the tonka bean collectors, the seasonal harvest is irregular, a job that must be carried out in the wild, and one that offers uncertain returns. What is more, the trees risk being cut down because of the quality of their timber. However, the tonka bean is held in such esteem by perfumers that for close to two centuries it has been—and continues to be—collected, dried and exported to Europe and to the United States.

*Left: Bags of fresh tonka beans (photograph courtesy of Pierre Ruch)*

*Below: On the banks of River Caura, home of the tonka bean trees in Venezuela (photograph courtesy of Jean Cuenod)*

Gilbert, my tonka bean supplier, had invited me to visit the forest and witness first-hand the challenges in keeping alive this harvest, a significant form of income for a secluded population. This Swiss national, originally from Geneva, with his Venezuelan wife Beatriz, started collecting beans at the turn of the century to assist the Panare and Piaora indigenous peoples in an area south of the Orinoco, a long way from anywhere. I had been buying beans from Gilbert for some years, and he would send me notes telling me about his business and what he considered to be the most significant challenges. He was working on trying to limit illegal deforestation in the area. In his view, the only real way of addressing the situation was to maintain regular levels of income in return for collection of the products from the forest. If trees were being cut down, it was as a last resort, a result of inadequate remuneration. He wanted our industry to understand the fundamental social significance of its tonka bean purchases, beyond simply preserving a historic ingredient in perfumery. The power wielded by the perfume industry's buyers and the use made of the product by those companies—which affected volume, price and compliance with contractual undertakings—also had a direct impact on thousands of families. The fragrance companies could not turn a blind eye to their role and responsibilities. In the early 2000s, this was a message attracting little attention; it would take ten years before the efforts of actors on the ground and shifts in consumer attitudes would result in a serious change in approach.

The tonka bean tree, *Dipterix punctata*, can be found throughout the entire northern region of the Amazon, the Guianas, in Brazil and in the Orinoco basin in Venezuela. As long as there is a profitable market for its beans, it will be spared being exploited for its dense, fine-grained wood. Dispersed throughout the forest, it is often found in small clearings well known to the tonka collectors. Following its attractive flush of purple flowers, in a good year it may produce up to seven thousand fruits that resemble kiwi fruit and hang from its branches at the end of a thin stem. Impatient parrots peck through the stems to eat the barely ripe flesh. At the heart of the fruit, a very hard nut conceals a kernel: the tonka bean. Light brown, smooth, and a couple of centimeters long, the bean has a sweet praline smell when cracked open. Its fragrance intensifies as it dries, a blend of praline, caramel and cut hay that it gets from a remarkable molecule: coumarin. Its name has become more familiar since the appearance of the grated bean in certain desserts. Sought after by pastry chefs for its aroma, which is more subtle yet stronger than praline, its name lends an appealing touch of exoticism to menus. In perfumery, the tonka bean is considered a must-have ingredient, often found in compositions with oriental notes, where it can enhance patchouli, vetiver or myrrh. It sits at the intersection of tobacco, honey, vanilla and benzoin, and has enough personality for fragrances to bear its name, such as Tonka by Réminiscence. From restaurant menus to the latest great perfumes, the tonka bean has taken pride of place among the gourmand spices.

We were on a three-day journey to meet the collectors of this kernel on the banks of the Caura, a river which, together with the Cuchivero, forms the historic heartland of the bean in Venezuela. Gilbert's business was based in Valencia, on the coast, the country's second largest city, from where we had taken his small aircraft for a two-hour flight due south to Maniapure, his family's base in the forest. There, Gilbert and Beatriz have built a very simple house in an idyllic setting, at the edge of a river of pools and waterfalls. Maniapure is situated at the center of their collecting zone, a fifty kilometer by fifty kilometer tract of forest, and Beatriz has established a bush hospital there which she runs with determined energy and unstinting devotion. Their son, Juan Jorge, is now carrying the torch of the family business, having become an expert in the tonka bean after many years and many harvests. He spent a good part of his childhood with the local indigenous people from the surrounding villages; his father has shown me photographs of him as a very young child, naked, his face decorated with red paint, sharing a meal of cassava and grilled squirrel with the children of the tribe. He spoke their language, and they considered him one of their own.

We were accompanied by two collectors in the dugout being confidently steered by Juan Jorge, each of whom was responsible for one of the collection areas, along with dozens of *sarrapieros*, the bean harvesters. The pickers may be indigenous or *criollos*, mixed-race Creoles who have settled in the forest villages. The *criollos* are often descendants of the renowned rubber tapper *seringueros*, impoverished laborers who were brutally exploited between 1870 and 1920, and sacrificed at the altar of the Brazilian rubber tree for the glory of the outrageous wealth of Manaus, the legendary and fleeting rubber capital. The fortunes of that city took an abrupt turn at the start of the twentieth century when the British brought rubber tree plants to Malaysia, where they thrived in enormous plantations, leaving thousands of tappers out of work and the Amazon basin deeply scarred. Juan Jorge employed fifty collectors and had three thousand pickers working for him. He and his father have patiently constructed this network over the years and now this man of the forest is his father's pride and joy. The son of a Swiss man who can almost call himself a member of the Panare . . . It's not a story you hear every day.

"Making my way down impenetrable Rivers . . ." The first line of Rimbaud's poem "Le Bateau ivre" came to mind as I let my hand drag through the water as the Caura's current carried the canoe along. Blending images of the riverbank with those from that famous poem lent this part of the Amazon another dimension: "The Rivers drew me on to where I wished to go." After half a day on the water, we came to a small beach marked with a few stakes and plastic bags; the head of a collection trail, it was where we would camp for the night. Juan Jorge grilled the fish he had caught on the way and, having taken a drink from the river, encouraged by Gilbert who assured us of the water's purity, we then washed ourselves.

Darkness fell quickly, as it does in the tropics, intensifying the screams of monkeys and birds. Before we took to our hammocks, Juan Jorge had a thousand things to tell me about this corner of Venezuela and collection of the beans. Here, as in Brazil and French Guiana, numerous communities have put down roots in the forest alongside the indigenous tribes, and have been struggling to survive for decades. Some of his pickers came from a village whose population was descended from slaves. All of these people are poor and living off very little; permits for the exploitation of softwood timber are difficult to secure, so collection of forest products is critical to their making a living. Tonka beans here, copahu balm elsewhere and, in the case of Juan Jorge's company, cinchona bark. Despite being settled, the inhabitants of these villages are still, first and foremost, hunter-gatherers. They track game in the forest, and collect nuts, beans, balsam and bark.

The next morning, we walked for two hours to meet up with a team of pickers at a fruit collection and storage point. Other paths led out from there, heading much deeper into the forest. A healthy specimen of tonka tree can produce twenty kilograms of beans in a good year. The Panare make their way through the forest carrying baskets of plaited palms, picking the fruits and bringing them back to clearings where they let them dry out before opening them.

Three Panare Indians were busy splitting the fruit and cracking the nuts. They were using a *mano de Piedra*, a traditional stone hammer that is buried at the foot of a tree at the end of the season from where it will be retrieved the following year. I was given a hammer and then joined the group. The stone head, precisely carved, felt good in the hand and its weight made it an effective tool. I managed to strike my fingers with all of my first blows, but the Indians earnestly corrected my technique and I succeeded finally in cracking open some nuts.

Our beans formed a gleaming, brown mound piled up in the sunshine, a veritable ode to abundance. One has only to break open the beans to smell the scent of their ivory- and purple-colored flesh. They are then loaded into baskets which can each carry seventy kilograms; the pickers load these onto their backs and carry them back to the river, which can be more than half a day's walk away.

While we were splitting the nuts, Gilbert explained to me how the local indigenous people would traditionally use the beans to flavor their tobacco and for medicinal purposes, primarily as an anti-coagulant and cardiac tonic. Collecting tonka beans has been a well-organized activity in Venezuela since 1870 and the beans have been a lucrative product for close to a century. When first brought to Europe, the product was usually packed into small barrels containing black beans covered in coumarin crystals, known as "frosted beans." The dealers with whom Gilbert did business when he was first starting out had told him that in order to avoid customs and tax blockades, they would hide the bags by dragging them behind their canoes in the water. After this time in the water, the fresh beans, which had been brown when picked, would blacken up as they dried out, and would be covered in white coumarin crystals. The process had been improved by soaking them in the local alcohol, namely rum. Beans were added to kegs of rum, where they would be left to macerate for two days and the barrel would then be tapped to recover the alcohol. The beans would "frost" over as they were transported and this became a mark of quality for buyers, a practice which lasted through to the 1970s. It is also said that the intention behind this custom was to make the beans unsuitable for germination, thereby preventing plants being grown in other parts of the world, the result of a lingering memory of the abduction of the Amazonian rubber tree to South East Asia.

But its medicinal properties would come to harm the bean's

reputation. Coumarin is suspected of causing lesions on the liver and lungs when consumed in high doses, and in 1954 the United States banned its use in flavorings, putting an end to its primary market, as a tobacco flavoring. The tonka bean then fell back on its other market, the fine fragrance industry, where regulation of its use means it is safe.

I ran into Juan Jorge again in Paris recently, fifteen years after my first trip up the Caura. After two decent harvests, he was keen to do the rounds of his customers in Europe and it was a pleasure to see him again. It was also an opportunity to speak on the telephone to his father, who had taken a step back from the business but was still just as charming. Time passes slowly in the Amazon. Nothing had really changed in terms of the way his business operated—neither its methods, nor its tools—apart from the arrival of G.P.S. Of course, Juan Jorge is now able to record the position of his trees more easily, but nuts are still cracked with a stone mallet. I was curious to know how the Panare Indian villages I had visited had changed, and if they were still loading the beans onto their backs to transport them. He told me that the only real progress had been the arrival of bicycles, which could now be bought for three or four days' wages. Malaria continues to wreak havoc and his mother's hospital has never been so busy, nor so in need of funds.

We spoke about his own harsh experiences of the effects of climate change which he had witnessed over the previous decade. He had had to manage the unprecedented situation of consecutive years of trees not flowering, and other years where the flowers produced only a meager quantity of fruit. He had been tempted more than once, in the face of such unpredictability, to walk away from it all. To my mind, it is that indigenous side to him, his connection to his Panare childhood friends, which has stopped him. Juan Jorge is also mindful that there are some benefits to be had from the tonka bean industry in Brazil. The Brazilian bean bears fruit in the Venezuelan off season, and it is this ability to balance and alternate between the two sources that has allowed the product to retain its position in the fine fragrance industry.

The continued existence of the places, people and techniques used to collect tonka beans flies in the face of history and the inexorable march toward modernity. Moving through the forests along paths that have been communicated from father to son in order to harvest kernels that need to be cracked by a carved stone mallet, before they are sold for use in a luxury industry, may seem improbable, indeed miraculous. How long can it continue? Returning from our foray into the forest, as I watched the trees pass by on the wild banks of the River Caura, I tried to imagine the future of perfume in the Amazon, whether trees would be felled or protected. It is a duality that reflects the hunt for an equilibrium that is critical to the people still living in and from the forest, dependent on the swirling eddies of creativity and luxury, silently alert to the signs that will tell them not to cut down the trees but to keep heading upriver and along the paths to the tonka beans.

# THE SACRED TREE

## *Sandalwood in India and Australia*

Sandalwood, agarwood and frankincense resin: these fundamental scents have such a deep and ancient lineage in perfumery that they have become almost mythical. I have saved my stories of these three trees, and their legendary fragrances, to conclude these accounts, three exceptional scents that have been associated with the religious, the sacred, the very essence of human existence. The potent curlicues of scent from these woods and their resin tell the story of the role of perfumes in ceremonies developed by humans to communicate with the divine since the dawn of time. A role that bears witness to the ability people have always had to discover and select the most extraordinary scents nature has to offer us. Unchanging, mysterious and precious, these products are testament to a spiritual, emotional and sensual journey that has endured for more than three thousand years, a golden thread that runs through the human condition.

*Kununurra, Western Australia: the stump is the richest part of the sandalwood tree in essential oil*

---

The first in this trilogy, and the subject of this chapter, is India's sacred tree, the sandalwood, a tree so venerated it was believed to be eternal before its own story turned to tragedy. One day when we were visiting his family's farm, not far from Coimbatore, Raja led me beyond his last rows of jasmine toward a large wooded area. "This is where we were growing our sandalwoods. They were already more than twenty years old . . ." It is a story which dates back to 2005, a decade before our conversation. About ten men, armed and masked, turned up one night at the houses occupied by the farm's caretakers and their families, declaring in a few short words, rifles in hand, that they had come to cut down the sandalwoods. They locked up the poor farmers, saying they wouldn't be hurt provided they made no attempt to leave their homes. The operation continued right through the night and the bandits left with the twenty-five trunks of sandalwood. Raja was scarred by these events. The savage felling of those trees was a violation suffered by his family, a sordid crime that had intruded into a unique source of beauty and spirituality for most Indians. I had already heard similar stories. Elsewhere, matters had often taken a turn for the worse and the sandalwood thieves had not hesitated to kill farmers who tried to defend their trees. In southern India, on several occasions, I had been shown the places where trees had been stolen and stumps ripped out. A few months prior to the theft, Raja had, like me, heard talk of the planting of thousands of hectares of sandalwood in Australia, a story causing ripples in the fragrance industry. "At first I reacted badly to the suggestion that another country would want to appropriate a tree that is part of our national heritage. But after what happened on the farm, I told myself I just had to accept it. We have inflicted so much damage on our own sandalwoods that perhaps we deserve their being brought back to life somewhere else."

Raja gave me a rundown of the bitter history of sandalwood's overexploitation in India. From the 1970s to the present day, this wood has endured everything from shortages to tragedy. Organized gangs have targeted the trees, stealing them in a display of mafia-like crime. The increasing scarcity of sandalwood trees, to the point where they have almost disappeared, is horrifying, and sits in stark contrast to the country's historical association with this wood and its fragrance, and its connection to the very soul of its people.

Sandalwood is deeply embedded in Indian culture. Rabindranath Tagore, winner of the Nobel Prize in Literature, said that his best prose and poetry were written after he had rubbed sandalwood oil into the soles of his feet, the palms of his hands and the crown of his head. "As if to prove that love will triumph over hate, the falling sandalwood spreads its scent, lending it to the metal of the axe that fells it," he wrote, reworking an ancient literary image.

Its striking smell, quite unique, at once woody and milky, is instantly recognizable. To Westerners it is bewitching, evoking a form of the sacred, of unadulterated exoticism and mysticism, and an irresistible vision of India. We spontaneously associate the smell of sandalwood with the smoke of incense sticks in Indian temples that we have either breathed in ourselves or conceived of in our imagination. Its fragrance is both loved and venerated by

Hindus, Buddhists and Muslims alike, in a convergence of appreciation which it shares with the oud that is obtained from agarwood. In India as well as in China, sandalwood is traditionally used in religious and other ceremonies, for medicinal and cosmetic purposes, and in artisanal handicrafts. It is burned, carved and ground to a powder or paste; there are so many facets to these timeless traditions. Buddhists burn sandalwood to accompany their prayers and meditation, Hindus use sandalwood paste to anoint the deities in their temples and the foreheads of pilgrims. The carving of this wood is limited to precious items such as prayer beads, jewelry boxes, effigies of Indian gods, and for special works of cabinetry, such as highly decorated palace doors.

As we meandered around the temple in Madurai, browsing in the shops where you can buy small sandalwood sculptures, Raja would often speak to me about his relationship with this fragrance. As a child, it was a part of his everyday life, both exotic and utterly familiar. In the *puja* prayer ceremony, in which an offering of flowers is made to a statue in the home, a little bit of sandalwood is grated and mixed into a paste with oil and camphor before being rubbed into the forehead between the eyes. "The *puja* sandalwood rituals left their mark on me before I was even old enough to go to school," Raja tells me, "and they are with us still at the end of our life. When we die, the purest way of helping the spirit of our deceased on their way is to add a piece of sandalwood to their cremation. In fact, it's considered very important, even if burning whole pieces is something only the very rich do. For me, sandalwood will always remind me of the scent of our soap, it's a quest for purity, a reminder of the divine in our life."

Sandalwood's long reign in Indian culture would not, however, be enough to protect it. A victim of its own success and of the popularity for the last century of its oil in perfumery, India's sandalwood has suffered no more enviable a fate than French Guiana's rosewood.

Sixteen identified species of sandalwood grow over a vast region of the Indo-Pacific, from India to Hawaii. Different oils, different scents, but all sharing the very recognizable "sandalwood" attributes. Even were one to include the four species endemic to Australia, as well as the sandalwoods specific to New Caledonia, Fiji, Tonga and Vanuatu, the king of the sandalwoods is still considered to be the white sandalwood, *Santalum album*, whose territory stretches from India to Timor, passing through Sri Lanka. The timeless tradition of using sandalwood for religious purposes and for woodcarving unsurprisingly saw a spectacular acceleration in the eighteenth and nineteenth centuries. In the Pacific, it was linked to the rapid expansion of the tea trade between China and the Australian colonies. The Chinese traveled the archipelagos in the region, trading sandalwood from the islands for cloth, metal, weapons and alcohol.

This trade in sandalwood would lead to both a significant disruption to local culture and a rapid depletion of sandalwood resources, as witnessed from the mid-nineteenth century.

In India, reserves of sandalwood were originally concentrated largely in the state of Karnataka, south of the city of Mysore, spanning a forested area of ten to forty kilometers in breadth by four hundred kilometers running north to south. It encompassed a vast expanse of mixed forest populations, as the parasitic sandalwood seeks nourishment from the roots of nearby trees. It is not a tall tree, it grows slowly, and the fragrance of its oil becomes more concentrated at its core with the passage of time. Over the years, the wood at the heart of the tree becomes very dense and takes on a beautiful, brown hue; the wood is rich in essential oil that lends it a persistent fragrance years after it has been felled.

By 1792, as it was increasingly exploited, growing ever more valuable, Tipu Sultan, the king of Mysore, declared it a "royal tree," claiming a monopoly in its trade. Appropriation of sandalwood by the Indian authorities, supported by the British, would continue to the present day, and be the primary cause of the tree's disappearance. Articles from 1910 provide a useful description of the large-scale organization of the industry, with two thousand tons of the wood being felled each year, classified by quality into eighteen different categories before being sold at auction.* Twenty years later, the annual yield would reach three thousand tons and the tree would be considered exploitable at just thirty years of age, rather than the traditional forty to fifty years.

Sandalwood's death warrant was signed by a decision of the authorities to claim exclusive rights to plant the tree, at the expense of any such right on the part of an individual. This monopoly put a stiff brake on ways in which the species might otherwise have been renewed. Replenishment of sandalwood stock would prove impossible. In 1916, the maharaja of Mysore built a large sandalwood distillery to help dispose of stock that had remained

* *Plantes à parfum* by Hubert Paul, 1909 appearing in H. Dunod & E. Pinat, Éditeurs, 1909, and *La Parfumerie moderne*, June 1910.

unsold while Europe was preoccupied by the world war. This initiative popularized the wood's essential oil, clearing the way for its triumphant return to the palette of perfumers. The oil, known as Mysore sandalwood, earned a reputation comparable to that of Bulgarian rose. Mysore sandalwood continued to be exploited to the tune of three thousand tons of wood a year through to 1960, despite every indication of its imminent depletion. Official figures for the year 2010 indicated a yield of forty-five tons of wood. Hardly any trees remain; they have reached the end of the road.

The level of corruption around the government monopoly in the trade in sandalwood is such that even today nobody knows the truth behind the figures. The industry has been slow to react, relying on generally dubious certificates of origin, for the wood itself as much as for the essential oil.

Gradually the industry turned its attention to Sri Lanka, another source of white sandalwood. But there, too, the resource is growing scarce and planting programs, while recently implemented, are limited. A shortage of Mysore sandalwood presents a real problem for perfumery because it is used in hundreds of formulae. The use of essential oils produced from different species of sandalwood, whether from New Caledonia or Australia, offers only a partial solution to issues of substitution and provides little comfort to perfumers who have had to do without the quality of Mysore sandalwood in their creations.

In the early 2000s, the first whispers of an intriguing initiative had the profession all ears, taking Raja by surprise, too. Would the establishment of thousands of hectares of plantations in the north-western Australian desert open up the possibility of an entirely new, prolific and sustainable source of *Santalum album*? My company had decided to stop purchasing sandalwood from India; the networks that remained lacked any real transparency, and things were only getting worse. As for our supply from Sri Lanka, it was insecure, limited, and corruption surrounding export quotas was rife. Sandalwood had become a major headache for me. There was a real risk of finding oneself implicated in violent crime. We spoke to Raja about it; he was regularly declining offers of batches of wood in return for cash payment. Any involvement with the sandalwood trade was making him nervous. Fifteen years after the plantation of the first trees in Australia, it was time to pay a visit to this new forest in the southern hemisphere.

It is a five-hour flight from Perth to Kununurra, a small town that feels like it may as well be at the end of the world, so hidden away is it in the immensity of Australia's north-west. And yet it is a town whose name deserves to be known. Right on its doorstep sits an enormous mine producing highly desirable pink diamonds, said to be the rarest, most precious diamonds in the world. Worth up to millions of dollars, these extraordinary gems are displayed in an understated, deserted shop in this sleepy town that cowers under the crushing heat. While I am there, I find out that the diamonds have become too scarce and the mine is going to close, that Kununurra will have to find fame elsewhere, and that sandalwood may just be it. Twenty years ago, Australian producers of an essential oil distilled from an endemic sandalwood species, *Santalum spicatum*, met up with investment funds specializing in fiscally attractive forestry investments. The notion of investing in Indian sandalwood plantations was considered and researched. And the idea took hold. The Kununurra region was chosen for its climate, its soil and the potential for unlimited irrigation from Australia's largest man-made lake, which is right next door. Today the project is up and running, despite some bumps along the way, and notwithstanding some rivalry and disputes between competing growers and investors. It has taken fifteen years for ten thousand hectares of sandalwood plantations to emerge from the earth.

My guide in Australia is a French engineer, Rémi, a Harvard graduate who had worked for several large American companies. A native Parisian, he had spent many years living in New York before taking a position with a large fragrance group in their acquisitions team. A few years ago, he took over management of one of the two companies that between them are responsible for the plantations, and he now finds himself managing three thousand hectares of sandalwood in Kununurra. It was he who suggested I pay a visit to his production projects. Rémi was already familiar with the world of fine fragrances; thanks to this new job, he has become an avid connoisseur of their raw materials. A jazz musician and horse-breeder, he has his roots in Paris and Brittany, but the adventure of developing a new model for sandalwood convinced him to accept for the time being a life of shuttling between France and Australia. With sparkling eyes, Rémi reflects on his new role with a touch of humor and unfeigned enthusiasm. Once a stalwart of cocktail parties and perfume launches, he tells me how he has employed and motivated a local team here in outback northern Australia. He has had to start from scratch, establishing nurseries and planting hundreds of hectares every year. He has now started exploiting the older wood, chipping it in order to send it by truck to the distillery in Perth, three thousand kilometers to the south, an exercise that matches not only the enormity of the country but the scale of the ambition involved in this risky venture. There is much at stake for both of us on this visit. I am keen to prove to our perfumers that these Kununurra trees, the oldest of which have been in the ground for only fifteen years, can provide them with an oil that matches the specifications of the Indian oil sufficiently well to replace it. "I'm not going to let you leave before I've convinced you!" Rémi says to me, very much aware of the inherent difficulty of the exercise.

As the afternoon draws to a close, he takes me to see his most recent plantations and I am stunned by the spectacle. Thousands of perfectly aligned little white sheaths adorn the vast, dark expanse of earth for as far as the eye can see. Small tubes protect the young seedlings, which are still showing only a few leaves. "It's a sixty-hectare plot," Rémi says. "We're testing a new spacing plan between the sandalwoods and the host trees." The agronomist in charge of the plantation has joined us. He is explaining to me precisely what Raja had just started to tell me. The roots of the sandalwood need to feed off the roots of nearby trees from another species, but not just any other species. The selection of these "host" varieties, their spacing, the amount of time they are left to grow alongside the sandalwood, all of it will determine whether or not the plantations are successful. Rémi confides in me that he has set himself two personal goals while overseeing this project: the first is to find the best method of growing the trees, and the second is to develop essential oils that can be used by the best perfumers from the wood of trees that are still young. The plantation before me is vast, on a completely different scale from the aromatic tree plantations I am accustomed to seeing. Rémi watches me, pleased with the effect he is having. "This year we will plant 240 hectares. What do you make of that?" There is a great deal of good to be made of it, as far as I'm concerned. My gaze takes in the young seedlings that stretch into the distance, and a memory of the devastated stand of trees on Raja's farm flashes back to me. I can feel a sense of excitement which makes me want to drop everything, pick up a spade and set to planting trees. The sight before me reflects every one of those stories of tragedy, theft, black-marketeering and depletion. The only thing visible here is renewal and growth. Night falls, the Tropic of Capricorn skies turn incandescent. "Look at those shades of pink, that's the color of the evening skies here. The locals say the sky is sapphire pink because they daren't go so far as to say diamond pink . . ." The sunset colors the protective tubes around the trees, and I find myself lost for words. Rémi has to drag me back to his car. I'm impressed by his dynamic energy. He finds it amusing that there are crocodiles infesting the local waterway and doesn't seem bothered, either, by Kununurra's crushing heat. "It doesn't take long to get used to it," he murmurs, as I collapse that evening, undone by jet lag and the sapphire sky.

The next day we visit plantations with a variety of trees of every age. Rémi is convinced they are about to find the optimal model for production. He has already made me promise to return in ten years' time to see the results. He has kept the best part of the visit for late in the afternoon: the plot with the oldest trees, planted fifteen years ago. It is still very warm. Rémi directs me toward the sound of some machinery at work; we're off to see the harvesting of trees that have reached an age where they can be distilled. It all comes down to this: are the trees already able to produce an essential oil acceptable to our perfumers? We walk through the plantation in the relief of the shade offered by rows of trees five to six meters tall, the sandalwoods and their host trees alternating in an attractive array of shapes and branches, creating vaulting archways that are starting to resemble a forest. Thinking back to Raja, I can't help feeling that this is what the landscape in the Indian hills of Karnataka should have looked like. Instead, everything that is being recreated here has already disappeared there.

The neighboring plot is in the process of being harvested, an excavator is hard at work. The sandalwoods are being uprooted so that their rootstocks, which are very rich in oils, can also be harvested. Rémi is walking ahead of me. We watch as the excavator digs up a tree and suddenly, with no warning, I am assailed by the warm, intense smell of sandalwood. "Come and see what you're smelling," my guide says to me. Some thick roots have been left behind in the hole from where the tree has just been pulled out, some of them sliced open by the excavator's shovel. I move closer; this is what smells so good. I pull one out, it is an orangey color, and purple on the inside. The familiar scent is even more aromatic in its natural setting, a strikingly potent, almost disconcerting blend of intense woodiness and delicately heady milkiness. I have kept that piece of root. Four years later, it is still releasing its scent.

---

The trees are washed in great drums on a platform before being stripped of their bark and chopped, then ranked according to quality. Of the eighteen traditional Indian grades referred to in industry articles from the beginning of the twentieth century, only six are still used. The profile of the essential oil is very sensitive to the various parts of the tree that are distilled: rootstock, lower and upper trunk, thicker and thinner branches. They each require separate distillation and blending processes in order to produce the quality required by fragrance creators. The biggest issue, and one which is constantly present when dealing with aromatic trees, is their age. In northern Australia, investors cannot afford to wait for the trees to reach the ideal age of forty to fifty years. A decision has instead been made to start exploiting the trees once they reach fifteen years' maturity. This seems to be the youngest end of the spectrum; no essential oil from younger wood has been considered acceptable. Trees never like to be rushed. Back in the sheds, the different categories of wood are separated out after being stripped of their bark, and Rémi reviews the possible blends as he discusses what may be suitable for us. We follow him into his brand-new, air-conditioned laboratory, surrounded by a team of white-coated Australian technicians. I'm hugely impressed. The set-up is not dissimilar to our Paris facilities, and a far cry from the rudimentary facilities of the sandalwood distilleries I've visited in Sri Lanka. Rémi takes great delight in finding similarities to familiar fragrances, brands and niche scents in this evaluation session. He has thoroughly analyzed my oil from Sri Lanka and explains how he is trying to reconstitute it by blending different fractions. We smell the ten or so samples he has prepared and the difference between them is marked. The young wood accentuates the milky characteristics of the sandalwood in the oil, which to my taste is a good thing, but the scent diverges from our standard. The best samples are those that comprise a significant fraction of root wood, and two of them seem to be getting very close to the goal. Our experts in Paris and Geneva will settle the matter. I smell the tester strips while still holding my piece of root, whose scent is irresistible. We're back out in the sun now, standing in the middle of the curing platform where stripped white trunks have been heaped into an enormous pile to dry out for a month before being chipped. This impressive heap reminds me of an image from the 1920s that Raja showed me: some Indian laborers had been photographed with their serious, turbaned and mustachioed foreman, posing in front of a heap of sandalwood ready to be sold. A century later, I ask Rémi to come and recreate the Indian scene for a photograph with me in front of his treasure, but this time sporting baseball caps and sunglasses.

Before leaving Kununurra, Rémi suggests I plant a tree. "Then you'll have to come back to see it. And it will give you a chance to walk in the shade of my plantations, which will have grown a lot by then. It won't be long, you'll see." An idea, some money, some water and a great deal of energy . . . That's how a forest of fragrant trees has emerged from this earth amid diamonds and crocodiles. It takes some considerable time for Rémi's samples to be analyzed in Geneva on my return from Australia. In the end, two of them are accepted and their blend will constitute our new standard.

Back in Paris, I decide to organize a meeting between Rémi and Raja. I tell them about the parallels between the fates of rosewood and sandalwood. In the twentieth century, two hundred thousand tons of wood from each species were felled, enough to bring them to the brink of extinction. It seems inconceivable, but a 4,000-year-old relationship between humans and sandalwood was almost wiped out in less than a century. Raja, the jasmine producer, has always refused to turn his hand to sandalwood distillation but he is intrigued by our Kununurra stories. When the conversation turns to India, he explains that the Indian government has been forced to react to the shift in the market caused by the Australian plantations. The legislation that regulates sandalwood is changing and it is once again possible to establish privately owned plantations. His uncle will now be able to exploit the fifteen hectares of trees he had fenced off and has been keeping under guard. He is no longer obliged to sell everything to the government. The historic monopoly that has caused so much devastation is finally coming to an end and Raja is hopeful that the sandalwood industry in his country can be revitalized, knowing, however, that it will take twenty or thirty years to see any significant results. "Frankly, without the appearance of Australian sandalwood, nothing would have changed, that much is clear," he says to Rémi.

It is a wonderful story. In twenty or so years, with its thousands of hectares of mature plantations, Kununurra will be able to bask in the pride of being the sandalwood capital of the world. A diamond of the fragrance industry will have symbolically taken over from the town's pink diamonds. I have promised myself I will return to see my tree when it is fifteen years old, at sunset beneath the sapphire pink sky.

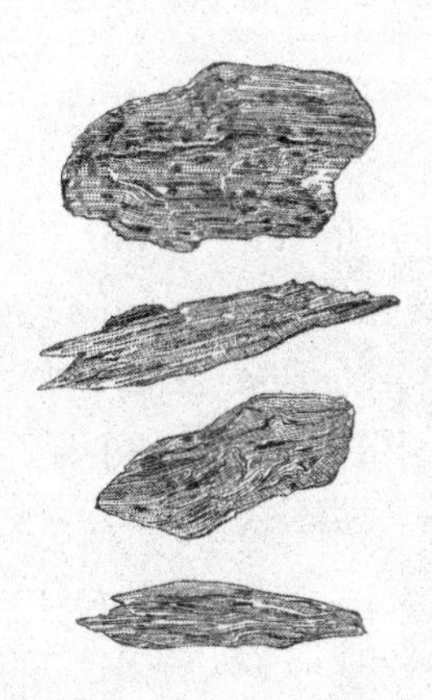

# THE WOOD OF KINGS

## *Oud in Bangladesh*

It was in the silent forests of Bangladesh, home to tigers and cobras, that I came to understand the real meaning of the name "wood of kings," thanks to the tales shared by the master craftsmen who have worked with this tree, and their accounts of its fragrance and history. Its wood has been known by many names from antiquity through to the sixteenth century: "agarwood" in Sanskrit, "aloeswood" in the Bible, "eaglewood" as used by Portuguese seafarers, and, to the Arab people, it is known simply as *oud*. It also goes by the name "wood of kings," the name which perhaps best reflects its value, its uniqueness, its potency; indeed the very aura it has generated throughout history, from the courts and palaces of India to Versailles.

All of these names describe the concretions formed by wood rich in resin which develops within trees of the *Aquilaria* genus.

*The guardian of the Adompour agarwood forest in Bangladesh*

As a way of defending itself against attacks by fungus carried by insects and inserted into the tree at the sites of any wounds or weak points, within its softer white the heartwood of the tree develops cores of a dark, dense wood generated by a resin whose olfactory potency is unparalleled in the plant kingdom. This alchemy is a remarkable phenomenon, resulting in a product that remains hidden within its trunk, like nuggets of gold in a river bed. There is something mysterious about this secretive mutation from white to black that produces the oud of the *Aquilaria* tree. It is a mystery which pervades its history and that of the people who look for it, work with it and use it. I have felt its mystique from the moment I arrived at its birthplace in Bangladesh.

Oud essential oil is fundamental to Middle Eastern perfumery. The Western world overlooked it until recently, but now it is rediscovering this sumptuous fragrance, whose strength and voluptuousness reflect the quintessence of this union of resin and wood. Men in the Middle East are fond of using a drop of oud to perfume themselves—indeed it is used or evoked in every Arab fragrance. And for the last decade or so, perfumers in Western cultures have been keen to procure it for use in their own compositions. At the start of 2015, I left on a trip to northern Bangladesh in the hope of seeing oud growing in its natural environment, deep in the historic region of Assam, an area that is now shared with India. It is not possible to travel on one's own to the place where it is sourced; an invitation is necessary. My guides were the heirs to a living but hidden tradition. Muslah, a Bangladeshi steeped in Sufism, and Damien, his French partner, welcomed me to their distillery in Susanagar, not far from the city of Sylhet. Over the course of various encounters, I had won the trust of these two childhood friends, who were quite dissimilar yet complemented each other perfectly. They had met by chance in Paris, where Muslah had family, and had not parted company since. When looking for a business partner in the global fragrance industry, they contacted me and convinced me of the extraordinary nature of their story. It was not long before they were insisting I come to Sylhet, where their company manages a million *Aquilaria* trees, a legacy created by Muslah's family over eight generations. Muslah and Damien are the oud kings.

There is magic at work, according to Muslah, in the relationship between Hussein, his master distiller, and his products. Hussein has carefully carried out his basin and has settled himself in the small distillery's sunny courtyard. Once a week, he carries out the racking procedure, which clarifies the oil resulting from distillation of the oud wood in his array of small stills. The metal receptacle he is using is the size of a bucket and it is almost full, its surface gleaming in the sun. Muslah is standing next to him, silent, watching my reactions. Together with Damien, we watch as Hussein puts down the bucket containing the water and oil, sits on his stool and cleans the steel vessel that he will use to collect the product of the distillation process. This simple traditional object in the shape of a saucer with a pouring beak is a critical tool in the process used to collect the rarest, most expensive raw material in perfumery. Costing between 30,000 and 50,000 U.S. dollars per liter, the best oud oil is worth five to six times more than rose oil. Hussein gently places the palm of his hand on the surface of the liquid in the bucket, as if just grazing the top of it, then lifts his hand up again and slowly scrapes the precious liquid left on his skin against the edge of the saucer. It is a method that has always been used in this part of the world, which is where the distiller himself was born. Hussein is imperturbable; his gestures, learned from his father, are precise, meticulous. He has been distilling for thirty years and he knows every nuance of the process required to obtain the quality that Muslah expects. The only sound to be heard is that of birds singing in the sunshine.

Dressed entirely in white, head covered, an upright, elegant figure in his traditional clothes, Muslah watches on as he talks to Hussein in a soft voice. He hails from a noble family of dignitaries, adherents of the Sufi faith, a religious and philosophical branch of Islam that has existed in the Sylhet region since the great saint Shah Jalal established Sufism in the area around the year 1300. Having been independent for a long time, the kingdom of Assam was coveted for its many herds of elephants and its agarwood, and in the seventeenth century it was conquered by the Mughals. Muslah's family has been planting, managing and exploiting these trees for their wood and for their oil for eight generations. The young man has inherited hundreds of hectares of *Aquilaria* plantations, trees of every age, some said to be up to two hundred years old. Such ancient *Aquilaria*, if they are indeed that old, become the stuff of legend, and must be so valuable that they ought to have been felled a long time ago. Muslah reigns over close to a million trees, a formidable resource that he manages according to one simple rule: remove as little as possible to ensure an even richer legacy is passed on to his son's generation. Guardian of his family's treasure, he is the king of Sylhet's trees.

Damien, whom Muslah considers a brother, has his own unusual story. He is very proud of his Circassian heritage from the north Caucasus, a part of the world with a history as rich as it is complex. He is interested in history, probably as a result of his own background, and his passion for oud was borne of his friendship with Muslah. He divides his time between France and Bangladesh and has compiled an enthralling work,* the result of years of research into the extraordinary fortunes of the agarwood tree. In it he gives a detailed account of how its resin produces a smoke that is associated with the divine in Hinduism, and with illumination in Buddhism. It was burned in the Temple in Jerusalem, it accompanied the Prophet Mohammed and it is thought to be one of the perfumes used to anoint the body of Jesus. The aromatic wood's reputation is such that it was considered by Christians in the Middle Ages to have its origins in earthly paradise, and it was depicted floating over the rivers that flowed out of that garden of earthly delights. Wherever there is oud to be found, magic is never far away. Revered in China and Japan, it is considered one of the three most important oils in the Arab-Muslim world, together with rose and ambergris. It is a jewel that has been coveted by the leaders of every Far Eastern court and every medieval king of Europe. Napoleon, we know, was fond of burning it to purify the air. Damien is one of the greatest connoisseurs of this substance and through his research and translations of ancient texts,

* *Le Bois du paradis* by Damien Schvartz, unpublished.

he has traced the path of this "wood of kings" through every culture. When it comes to the history of oud, he is royalty.

Far from Sylhet, Alberto rules over a very different kingdom. For more than three decades, Alberto has been a celebrated master perfumer. He is uniformly acclaimed by his peers, the highly decorated creator of countless successful fragrances, and is sought after by the biggest perfume houses, ever inventive, relentlessly hardworking and never still for a moment. It is not something that Alberto himself will ever concede but everybody acknowledges it: the boy whom his father would call "my king" has himself become king of his generation of perfumers and a quite extraordinary character. With a childlike simplicity and authenticity, Alberto hunts down the beauty in all things and in every moment without ever losing his sense of humor. He either *loves* something or *detests* it, and he will declare his opinions to anybody who cares to listen. With his pale blue eyes and intense gaze, he has the natural elegance of a "Don" from his home town of Seville. He has made a life in Switzerland, where he is passionate about tending his garden, a masterpiece of white flowers. Alberto sees something fluid, something elusive in his creations. His perfumes seduce and endure: Must by Cartier, Pleasures by Estée Lauder, Flower by Kenzo, Acqua di Giò by Armani, CK One by Calvin Klein, Bloom by Gucci, all of them have become true classics. Dozens of others bear the mark of his strength and delicacy, as he plays with the most beautiful of natural ingredients with both conviction and subtlety. It was imperative that Alberto be introduced to oud, and I was fortunate enough to be involved in the encounter between these two men, each kings in their own right searching for one another before their paths finally crossed.

The story dates back to 2015, and my first trip to Bangladesh. On my return, I had given some samples of oil to our master perfumers. A few months earlier, Alberto had made us all laugh when he declared, "Oud, oud is like a unicorn . . . Everybody talks about it but nobody has ever seen it!" It was a good-humored reference to the opacity of the oud industry, with countless products bearing its name despite being of uncertain origin and having been cut with synthetics. It was a labyrinth in which many a perfumer would lose themselves. Oils would be sold as oud at price points varying from 300 to 30,000 U.S. dollars per kilogram, and nobody had any idea what they were dealing with anymore. I was aware of the problem, and I knew that Muslah and Damien's oud would work its magic on Alberto. When he smelled a scent strip of their top-of-the-range oil that I had brought back from Sylhet, he said nothing, nodded his head and went straight to his iPad, a sign that he had been won over, and that he already knew where and how he wanted to use it.

A year later, we met up again for the launch of Man Wood Essence, a superb masculine fragrance that Alberto had developed for Bulgari around the image of an imaginary tree with vetiver roots, cedarwood trunk and branches, cypress foliage, all linked by its sap, which was represented by copahu balm. We were talking about the intensity of the woody notes in the fragrance and I asked Alberto if he had been making any progress with the oud. "Well, I'm in the process of using your oil in a whole collection. It's working to enhance my other naturals—I'm very pleased with the result." In one year's time, Alberto, this perfumer who had already been recognized with every possible honor in his profession, would also become the king of oud.

On my second trip to Sylhet, Muslah and Damien had me visit one of their properties by the name of Adompour, a forest that was the result of decades of *Aquilaria* plantings. The rice fields around Sylhet are broken up by vast forested hills which the local people refer to as jungle. The greatest expanses of forest are still home to tigers and king cobras, whose nests can sometimes be spotted burrowed into the hillsides. The jungle is a mysterious landscape inhabited by a variety of ethnic groups, the result of waves of peoples who have passed through this migratory crossroads. Some villages are Buddhist, others Christian; all of them live peacefully among the majority Muslim Sufists while remaining fiercely attached to their traditions. In these quiet woods, the only sound to be heard was our footsteps on the thick layer of dead leaves as we made our way through trees of every age. One can spot the *Aquilaria* by the light-colored bark of their smooth trunk patterned with yellow, green and brownish-orange lichen and fungi. Originally planted in rows, and aged from ten to eighty years old, they had merged over the years to form a mature stand of trees. In some places, a group of trees looked quite peculiar, revealing rows of sharp metallic tips which turned out to be nails. Ever since the eighteenth century, the people of Sylhet have been using a tree-spiking technique to speed up the formation of oud inside the trunk. When the tree is between fifteen and twenty-five years old, lines of nails spaced ten centimeters apart are driven into the trunks, almost entirely covering the tree. At first the nails cause the tree to sparkle in the sunshine, then they rust and are swallowed up by the tree itself as it grows. Each wound caused by the nails will be a possible site for the formation of a nodule rich in oud resin.

After walking for an hour, we arrive at a single tree growing apart from any others on the side of a hill. A generous crown of branches spreads from its impressive trunk. "There you go," Muslah says. "See, we weren't making things up, this is the one, he's real!" "He" is an *Aquilaria* tree which, as legend in the nearby village would have it, is two hundred and fifty years old. I savor the rarity of the moment as we head over to take a closer look. My gaze slides upward, following its enormous branches and the outline of its expansive crown, a tranquil yet complex composition of hundreds of thousands of leaves. If we look at them for long enough, these ancient trees will speak to us. They know stories dating back to before we were born, and they know what will become of the world after we have gone. It is as if they are immortal; some of them surely must be. There are scarcely any throughout all of South East Asia that compare to the one in front of me. Any tree of that age is so valuable that it is cut down as soon as it is discovered. The older the tree, the more likely it is to be harboring unusual pieces of wood within. It is said there are pieces of oud wood worth, in some cases, up to several million dollars on the Chinese market, depending on their size, the beauty of their shape and their density. Over the centuries, *Aquilaria* trees have been tracked down throughout Asia by hunter-gatherers who jealously guard their knowledge. Becoming first rare, then endangered, they, like rosewood, now feature on the list of protected species. The growing appetite for this wood has resulted in plantations becoming more widespread over the last fifteen years, and they now grow from India to Vietnam. Millions of young trees are subject to every sort of experimental treatment, such as injection with chemical products in order to encourage the formation of the precious resinous wood from when trees are only ten years old. But there is no sign of any of this here in Bangladesh. It is something of a rarity to see the hundreds of old trees which have been preserved in the areas afforested by Muslah and Damien. "We will never cut down this one. And, should we decide to fell it, we'll let you know and you can come and do it with us," Muslah says, laughing. How was it possible that this *Aquilaria* was still alive, that it hadn't been the target of poachers? Muslah tells me that the forest is guarded by villagers. "We can't see them, but they're watching us, following us. We have such a strong tradition of respect for the trees that any black marketeer would be noticed before they could get close. If they so much as dared to touch a tree, they would risk being killed on the spot."

We have resumed our walk when suddenly a figure appears at a bend in the track. An older fellow, thin, bright-eyed, his beard dyed orange, is watching us approach. He is standing perfectly still, a large stick clutched in one hand and his machete in the other. Muslah goes ahead to speak with this guardian of the forest. The man is a veteran of the 1971 Bangladeshi war of independence. He is barefoot and has scars on his legs, a souvenir, he says, of an encounter with a tiger. He knows the trees, and points us to the most promising of them, tapping their trunks with his stick. "Are you going to the village?" he asks as we finish. "You can carry on by foot, the tiger isn't here."

A little further on, as if by magic, four little girls materialize on our path. They seem to have appeared from nowhere, and are staring at us out of dark eyes. In the shadowy forest they are a riot of color, dresses in yellow, pink and white-flowered red, trousers in parrot-green and turquoise. As we draw closer, they disappear in a flash; it is as if they have melted back into the trees. Smiling yet serious, Muslah murmurs that we have just glimpsed the forest fairies. We catch sight of them again as we near the village; just a glimpse of a smile and they are gone. Silently and gracefully they appear and disappear, fairies casting a spell as they lead us to their glen. There they are, waiting for us, as we reach the first houses, as if granting us permission to visit, each of them perched on the branch of an old tree, splashes of color in the foliage, little dark-haired fairies gazing at us intently with curious eyes and grave faces.

The next day we set off to fell a tree which is likely to contain some pieces of oud. Damien and Muslah have marked the trees to be chopped down that season, knowing that it takes thirty to forty trees to gather enough resin-rich wood for a single kilogram of oil. After my first visit, we had agreed that they would reserve their oil for us and I had confirmed that our requirements for that year had increased. We were joking about the fact that their oud was too good, but it was a serious subject. They have set a limit on their production in line with their resource management plan that they do not intend to exceed, regardless.

At the end of the seventeenth century, in order to secure the future of the French royal navy, Colbert planted in the middle of France forests of oak trees that still exist today; in the same period, Muslah's ancestors were embedding the first nails into trees around Sylhet. I'm enchanted by this parallel: it speaks to the nobility of age-old practices of true silviculture, and to the power of the counsel given by the wise advisers to both the Sun King and the Mughal emperors. That day, the villagers tell us that the chosen tree is sixty years old, its rusted nails barely visible in its bark. After roping up the twenty-meter tree to be sure of clearing the nearest house, the two woodchoppers set to work. It is a choreographed display, their two axes working in precise, economical actions, and the tree falls quickly. At its base, an even distribution of black marks through the white wood indicates the richness of the tree's infection. The trunk is then split with wedges; axes cannot be used for fear of damaging the precious pieces of oud that have formed in various shapes which are not always easy to detect. Our tree is split into quarters, revealing the black nodules clustered around what remains of the nails. It is quite rich in resin, and Muslah has already noted a few concretions that would warrant being kept in pieces of wood. When placed on a bed of coals, any resin in the wood warms up and, once it starts to burn, it produces the famous smoke which Buddha is said to have described as evoking Nirvana.

On our return to Susanagar, we are shown around the carving workshop by Hussein, the distiller. In a vast, shadowy room, twenty or so young men are sitting cross-legged with a wooden chopping block in front of them, using their machetes to split off pieces of the *Aquilaria*'s trunk. Any wood which is not completely white will be used to produce the essential oil. The men are trained for a year to hone the precision of their cutting technique. The results of their labor are then meticulously sorted into separate baskets. In the next room, craftsmen are busy chiseling away at tortuously shaped pieces of wood to remove any remaining white wood. It may take four or five hours to finish a single piece, a whole day for the more difficult pieces. Here in the workshops, Hussein is ruler of his domain, supervising with a vigilant eye. He is uncompromising when it comes to quality. The pieces must be perfect in order to earn their full value.

Night has fallen, and the dull sound of machetes on chopping blocks has come to a halt. In front of the distillery, Hussein has set out a table and a few chairs. Muslah is keen for me to smell some different grades of oil. He stores his productions in square glass flasks that resemble bottles used for spirits. Traditionally, these oils should be smelled against the skin. Muslah puts a drop on the inside of my wrist, rubs it gently and leaves me to breathe it in. The scent of this dark, indecipherable woody fragrance blended with animalic warmth is always a shock. The raw leather notes in oud oil are astonishing, perplexing even. This woody blend is unique. For some, the "nanny goat" aspect to the product, evoking a stable or cheese, is off-putting. For most perfumers, however, it represents a pinnacle.

Damien has prepared a small container filled with coals, over which he rests a piece of black wood. The heat makes the resin explode in little bubbles as it comes to a boil, well before the wood itself starts to burn. I close my eyes as we're enveloped by smoke.

Whether captured in wisps of smoke or oil, the smell of oud is strong enough to make you swoon, to lose your bearings. Pure oud transports the mind, only reluctantly releasing its grip, it goes to your head like some form of legal opiate. As night falls in Susanagar, the potency of the scent brings images from the day dancing before me. The immortal tree, the guardian of the forest and his tigers, the fairies with their shining eyes, the splinters of dark and white wood, the deep brown of the oil on Hussein's hand dripping gently over the curve of polished metal gleaming in the sunshine.

In due course we presented Muslah's oil to our perfumers, as well as giving selected clients an opportunity to smell it. At the end of winter 2018, I was accompanied by two intrepid women from the fragrance division of a large jewelry brand on a trip to Bangladesh. They had loved listening to my stories and dreamed of bringing back their own memories of glinting nails, fairies, forest guardians and narcotic smoke. In May, they would be launching a collection of four extraordinary fragrances created by Alberto, each based on oud's majestic notes.

The launch party was to take place during the Cannes Film Festival, on the terrace of a grand hotel on the promenade de la Croisette. The contrast between the forests of Sylhet and the spangled glamour of Cannes could not have been more extreme. Like all great perfumers, Alberto is used to such shows. The launch of any new scent invariably means making himself available to journalists in elegant—sumptuous even—settings since the brand is always keen to ensure that the new fragrance is prestigiously positioned. I had been invited on this particular occasion to talk about the unique story of the oud he had used in his compositions. Still carried away by their experiences in Sylhet, my traveling companions wanted the noble character of the raw ingredient and its artisans to be given pride of place at the evening's event, alongside the perfumes themselves. They had been profoundly affected by our trip, and had brought back superb photographs as well as a film. Damien was also invited, and, while he enjoyed the evening and the guests, he was like a fish out of water. The guest list drew from the worlds of fine fragrance, jewelry, cinema and television. Everybody knew each other and there was much mutual congratulation. Alberto was in demand and made his way from group to group, entirely at ease, champagne coupe in hand. His sensitivity, his imagination and his experience make him a master of communication; he has a very appealing way with words. He called me over so I could share one of my stories with a journalist curious about the forest and the manner in which the oil is collected by hand. Alberto was listening to me, a smile on his face, and I knew that in his mind my stories were already transforming into scents. Damien had brought some pieces of wood along with him and as he started heating them up over the coals, the first curlicues of smoke floated out into the evening. Now images from a short film about Sylhet were being projected onto screens. The oud was finding its place among the evening gowns and champagne.

When the muse who had been chosen by the brand as the face of this new collection of perfumes appeared on the terrace, all eyes turned to her. A major Bollywood star, Sonam was radiant in a bright-yellow couture gown. With her beauty, charm, smile and penetrating gaze, she was aglow. The intensity of her presence led me to ponder the beginnings of an unlikely but undeniable connection between Bangladesh and this evening in Cannes. She stood for a moment, watching the projected images. On screen, Muslah, dressed head to toe in white, was observing Hussein's hands at work, and the orange-bearded forest guardian stared down the camera's lens with his intense gaze. Sonam made her way over to us. She was keen to understand the origins of the scent given off by Damien's smoke and to experience Alberto's four fragrances. After the introductions between Indian actor and master perfumer, the flacons were opened. She breathed in slowly, closing her eyes, then gently put down three of the scent strips and remained silent. Alberto explained each of his compositions, talking about his inspirations and the marriage of fresh notes with the intensity of the oud.

Sonam then asked me to tell her about Sylhet while she brought the smoking oud up to her face with a slow, gracious gesture. She was still holding the fourth scent strip in her hand, unable to relinquish it. She turned back to Alberto. "This one is extraordinary, how have you managed to create such a miraculous thing?" she asked, her dark gaze fixed on his own blue eyes.

Alberto was eager to respond. "Yes, I'm particularly fond of this one. I called it Nuit des Rois. It is meant to evoke images of princes in the golden light of the Arabian deserts in homage to the magic of the most beautiful of raw ingredients." He went on to list the notes he had brought together in the fragrance, splendid satellites spinning around the star that was the "wood of kings": bergamot, rose, patchouli, benzoin, sandalwood, cistus and frankincense. In his hand he held the flacon; it was an extraordinary thing to have brought together so many great natural oils into a single juice. And then suddenly, as I looked at the actor in her gown, which seemed to be emanating sunshine, it was as if I had in front of me an older incarnation of those little forest fairies, a sister from another world.

The enumeration of the ingredients used by Alberto set a reel of images running through my mind, like a private film projecting my memories of all the places I had visited over the years. Thirty years of traveling seemed somehow to kaleidoscope before me. It was as if Nuit des Rois were telling *my* story, capturing the many different fragments of my life in its heavy, blue flacon.

The yellow-dressed spirit had perfumed herself with Nuit des Rois, and Alberto's compositions wafted through the air, circling the champagne glasses, blending with the wisps of Damien's smoke as it came and went among the guests. Seeing Alberto surrounded by people, laughing, voluble and charming, his high spirits brought to mind a phrase I have always loved: "A king with no entertainment is a miserable fellow indeed." I murmured the words to him, thinking of Giono's novel of a similar name. Amused, he replied, "Oh, you know, I've been entertaining myself every day for more than thirty years. I would not have made a good king!" That evening, King Alberto had us all wanting to share in the entertainment, a celebration of a great artist's encounter with one of the world's natural wonders: oud, the wood of kings.

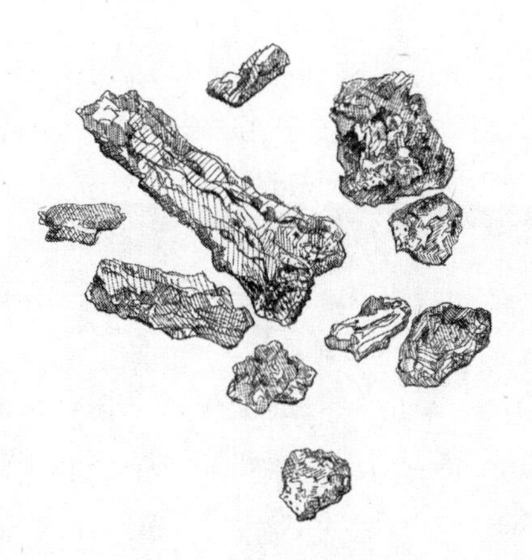

# WHERE TIME STANDS STILL

## *The frankincense of Somaliland*

> "Jerusalem . . .
> The multitude of camels shall cover your land,
> The dromedaries of Midian and Ephah;
> All those from Sheba shall come;
> They shall bring gold and incense . . ."
>
> Isaiah 60:6

I had been dreaming about traveling to the land of frankincense for twenty years. I had often imagined it as a final destination, the culmination of a journey that started with cistus in Andalusia. One day, I knew I would need to go and breathe in the scent of frankincense on its tree. It would be an experience that would give meaning to my exploration of the world's arborescent fragrances. I waited years to find the person who would make this

*The author in Hargeisa, Somaliland. Frankincense is sorted by female villagers into five different grades.*

journey possible. Then it took months to organize my trip to the Somalian coast on the Horn of Africa. The region has a dangerous reputation; large corporations fear for the security of their associates in this part of the world. When at last I arrived in Somaliland, I found myself already overwhelmed by a whirlwind of thoughts in which the footprints of mythical protagonists in the resin world from thousands of years ago blurred with those of the poet Rimbaud turned incense trader, and I had a sense of coming full circle after the three decades I had spent hunting for fragrances. The Egyptian queen Hatshepsut, Solomon and the Queen of Sheba, the three Magi . . . All of them joined the poet in a narrative that had always fascinated me.

Arthur Rimbaud spent the last eleven years of his life, from 1880 to 1891, between Aden in Arabia and Harar in Ethiopia. At the time, a traveler's itinerary between these two cities would start by heading to Zeila, a remote township on the African coast, from where one would join a caravan to embark upon a fifteen-to-twenty-day march. Ancient and mysterious Harar is one of the four holy cities of Islam. In the year 1880, it had been only thirty years since the first European had set foot there. For merchants operating in Aden, the opening up of a trading post in Harar was initially justified by the access it would allow to the coffee, or *moka,* harvests in the Ethiopian highlands, but it was also a gateway to many other products in the hinterland. As the manager of a desk in Harar, Rimbaud had a commercial seal made for him in the name of Abdo Rimb, meaning "incense trader." "I buy many other things, too," he wrote to his mother, "resins, frankincense, ostrich feathers, ivory, tanned hides, cloves, etc." An extraordinary inventory of items given that it had not, in essence, changed since Queen Hatshepsut's own journey twenty-four centuries earlier to Punt, which these days is assumed to have been located in Eritrea at the southern end of the Red Sea. Ethiopian frankincense was one of the items transported by the ten or so caravans that Rimbaud formed and led from Harar to Zeila. At the end of the nineteenth century, an era not so long ago, these caravans were enormously risky affairs, with frequent attacks costing the lives of traders and missionaries as well as those of French, Italian and Greek explorers. It was a period of history with which I was familiar, and I had brought with me *Rimbaud, l'Africain** on my journey to the land of frankincense.

More than mere fragrances, myrrh and frankincense are portals to the most ancient of histories, the stuff of legend, carrying with them the fantastical traces of caravans and civilizations that have long since disappeared. And on this journey—a journey so long as to evade our usual points of reference—perfume itself becomes a timeless marker.

When Queen Hatshepsut equipped a fleet of vessels to set sail for Punt in 1500 B.C., traders had already been plying the maritime routes between Egypt, the Horn of Africa and Arabia for close to a thousand years, an unimaginably long time. Trade focused largely on ivory, gold and feline pelts, but also, and perhaps most importantly, on resins, namely myrrh and frankincense.

A quest for the origins of frankincense takes one back through

* *Rimbaud l'Africain* by Claude Jeancolas, Éditions Textuel, 2014.

the mists of time; it has perfumed four millennia of human history throughout Africa, Arabia, Mesopotamia and the eastern Mediterranean. In ancient Egypt, myrrh was indispensable in the embalming process, and formed part of the preparations for the journey to the afterlife. Frankincense drops, which are drier, lent themselves to use as a fumigant.

The powerful scent of their smoke swirled around sacrificial altars and tributes to the gods. The source of these resins has not changed with the passage of time. Trees producing myrrh, of the genus *Commiphora*, and those producing incense, of the genus *Boswellia*, grow across an area that extends from western Oman and northern Yemen to Eritrea and Ethiopia, encompassing the entire northern coast of the Horn of Africa, and reaching down as far as Kenya. Myrrh and frankincense are gum resins, natural secretions produced by the trees when they are notched, and are part soluble in water or alcohol. And most of the frankincense used in the fine fragrance industry comes from Somaliland, a country that does not exist . . .

Formerly British Somaliland, the area now known as Somaliland has been awaiting recognition by the United Nations since its declaration of independence in 1991. No longer a part of the officially recognized Somalia following a war of secession, this land in the Horn of Africa looks across to Yemen and now shares a border with the Somalian province of Puntland, which is renowned for its maritime piracy . . . People are as wary of Somaliland as they are of its neighbors, hence its significant isolation.

---

I am met by brother and sister Guelle and Zahra at the tiny airport of Hargeisa, the capital of Somaliland, which has connections only through Dubai and Addis Ababa. They have invited me to visit them and will act as my calling card in this very inaccessible region. For ten years, Zahra and her brother have been patiently building up their business. They are pioneers in this part of the world in the promotion of ethical sourcing, and are as transparent as complex local realities permit. I buy from them myrrh and frankincense resins, which they are responsible for collecting, sorting and exporting. Their agreement to host me is no ordinary matter, rather it is a meaningful gesture of trust. They guide me through customs, making sure my passport is not stamped by the authorities, as any record of time spent in their country would make future entry into the United States very difficult.

As one leaves the city, asphalt soon turns to sand and modern buildings are quickly replaced by traditional Somali housing. The city center is not much more than a few low-slung buildings around a marketplace, and a hotel catering to businessmen from the United Arab Emirates and the staff of N.G.O.s. Hargeisa has practically no water supply, most of the city being supplied by water tankers. The dust-filled, sandy streets are lined with tunnel-shaped tents, structures made from wooden hoops covered by a colored patchwork of fabric, plastic bags, pieces of scrap metal and blankets. It is as if by persisting with such makeshift structures, these nomadic people are resisting the pull to permanent settlement, even here in the city.

The Somalis in Hargeisa, a tall, thin people, go barefoot. The women wear brightly colored robes and are covered and veiled.

The children appear alarmed to see white visitors in their midst—still a rare occurrence. Such is my welcome to a country reliant on breeding herds of camels and flocks of sheep for export to neighboring Arab countries, a country that lives mainly off the funds of the vast, active network of its diaspora. But the people of Somaliland are also involved in an activity that is deeply anchored in their history, namely the production and export of their resins to the world.

Driven by their values and belief in traceability and ethics, Zahra and Guelle were at first disconcerted by the mistrust directed at their country. But these days, the Western market is prepared to pay top dollar to secure supply from a trusted and sustainable source, a business that respects the players in the collecting chain, even if there is still considerable confusion between Somaliland and Somalia. It has required a great deal of patience on the part of Zahra and her brother, and there is still a long path ahead. Guelle spends his time between Hargeisa and Dubai. He is responsible for collection and production. Zahra is based in Europe, where she sells their product, and their story.

When I first met Zahra in Paris, she looked like the Queen of Sheba with her stately, princess-like demeanor. She had studied in Djibouti, before spending a number of years working for N.G.O.s in refugee camps in war-torn parts of the world. Burundi, Sierra Leone and Bosnia: Zahra has seen human distress in all its manifestations when following these conflicts. Finding herself in Rwanda in the immediate aftermath of the genocide, she feared for her life, and she resolved then to dedicate herself to development in her own country. Her voice is poised and calm, and her experience of the most challenging of circumstances has given her strength and determination.

In the accounts that tell of the mythical encounters with King Solomon not long after the year 1000 B.C., there is one image that stands out, whether true or not: an image of the Queen of Sheba, a dark-skinned Ethiopian sovereign, whose power of seduction over the king of Jerusalem was enhanced by the mystery surrounding her country's treasures. Zahra was obviously familiar with the story. Even if it made her smile, she was fond of reminding me of the primordial significance of resins, known as spices, in the encounter between the famous couple:

> *Then she gave the king one hundred and twenty talents of gold,*
> *spices in great quantity, and precious stones.*
> *There never again came such abundance of spices as*
> *the Queen of Sheba gave to King Solomon.*
>
> Kings, 10:2–10

Before setting off down the road that would take us to the legendary trees, Zahra and Guelle wanted to show me the port of Berbera, from where they export their gums. We had driven for four hours along one of the two sealed roads in the country in order to reach this port, which provides Somaliland with a gateway to the world. There are still some old streets around the fishing port, where handsome old Arab houses are gradually going to ruin. Guelle is anxious, the atmosphere at the entrance to the port precinct is very tense. We are trying to reach a boat that is being loaded with their gums, but faces are hostile at the entrance gate. Somebody points to me and the tone grows sharper. They are concerned about the role of foreigners in proposed changes to the port. Ethiopia, Somaliland's large, dynamic and ambitious neighbor, is looking for a passage to the sea, having been deprived of maritime access since Eritrea's independence, and it has settled on Berbera, with its links to the Emirates. The Chinese are funding the construction of a new road from Addis Ababa via Hargeisa to Berbera, and the port is in the process of being leased to the U.A.E. in return for significant levels of investment. Every Westerner is seen as a possible expert who intends to axe local jobs. The greatest aggression is displayed by young men under the influence of khat, a traditional drug whose leaves have amphetamine properties. A large proportion of unemployed youth chew khat all day long, which incites aggressive behavior in some. A group of young people approach me, shouting angrily. Guelle asks Zahra to take me quickly back to the car. He negotiates with them for an hour before they finally allow us through. Moored at the docks alongside a few old cargo vessels are some fishing boats. Next to one of them, workers are piling bags of Guelle's frankincense and myrrh into a container. Plodding up and down the adjacent beach are some camel trains about to be shipped off to Saudi Arabia. The scene playing out at the port of Berbera in the late afternoon is reminiscent of a bygone era of resin-laden caravans and dhows, as bags are loaded that will transit through Djibouti or Dubai before traveling to Marseille, and dromedaries walk slowly in single file along the beachfront.

---

From the earliest times, production and collection of frankincense have been divided between the Arabian and African coasts. It did not take long for a sort of maritime route to be established between the two shores in the form of rafts supported by inflated goatskins. Resins harvested in Africa would end up at the ports of what is present-day Yemen, from where they would head off on their overland routes.

Around the year 1000 B.C., the revolution that was the domestication of the camel led to the emergence of caravans. For ten centuries, the overland spice route would replace the vessels plying the Red Sea. By the seventh century A.D., there was a well-established route to bring frankincense into Egypt. Caravans would form for the sixty-five-to-seventy-day march that left Shabwa, Yemen's largest port, moving at a pace of thirty to forty kilometers a day, following the western coast of Arabia via Medina and then heading on to Petra and Gaza. Petra was a major trading crossroads of at least twenty thousand inhabitants, the meeting place for every caravan route from the south, and a redistribution point for goods being directed to the west, north and east. Resins constituted the principal source of wealth underpinning these trade routes. Caravans of frankincense were responsible for the emergence of the glittering civilization of the Nabataeans in Petra, before further significant developments brought about its decline. In the first century A.D., an understanding of the Indian monsoon winds was a significant development. It was now possible to sail east on a direct route to the spice coast of Kerala in southern India, then head up to what would become known as

Pondicherry. Ancient Greeks and Romans hastened to ply the waters of this new route for transporting spices, woods and resins from the East. On the return route, they would stop over in Arabia to load up with myrrh and frankincense. This highly profitable trade marked the start of the decline of the use of caravans. And the beginning of another story.

Returning from Hargeisa, Guelle has to finish organizing our expedition out to the trees. We need official permits and the police must appoint two armed guards who will accompany us, to protect us as well as likely keep an eye on us. While all this is being settled, Guelle shows me around his brand-new little distillery. A beautiful deep-yellow oil is flowing through his gleaming Indian stills. His eyes shine with pride as he shows off what, until recently, seemed unthinkable: industrial reform in a country almost devoid of such possibilities. Here, one must start at the beginning, build everything from the ground up. There is no infrastructure. Only three roads and a network of tracks, no higher education, no technical experts, no engineers.

I visit the warehouse next door to the stills. The building is full of bags of frankincense received from collection centers all over the country, bags that are packed with "raw" gum harvested from the trees. They now need to sort the very finest teardrops, which are translucent or pale yellow, from the second grade of more deeply colored pieces, and separate the smaller, greyer clusters from the pieces of bark. A young woman is overseeing the decisions made in this part of the sorting process. She has taken a big tray to demonstrate, and in a few minutes, with precise, rapid gestures, she has transformed a pile of loose pieces into little homogeneous batches. Zahra explains to me that they have had to teach this woman how to read and write; most often, the training of their employees starts with basic schooling.

The phenomenal success enjoyed by the aromatherapy industry in the United States since 2010 has resulted in a sharp increase in demand for frankincense. Both in Somalian Puntland and Somaliland, this explosion has prompted serious upheavals in the industry. Within the space of a few years, we have witnessed the disintegration of traditional networks and the uncontrolled and excessive tapping of those trees that are most accessible. There are more and more warnings in the media, descriptions of frankincense as an endangered product, the result of overexploitation of the trees.

Guelle prefers to smile at the situation. He is resigned; none of this agitated talk matches his own experience on the ground. There are immense expanses of trees in Somaliland. Myrrh can be found everywhere in vast areas throughout the country, and he will take us to see how he is opening up new collection areas, where he is teaching villagers how to tap trees that have never previously been tapped. Frankincense is a more complex proposition. The trees grow best at altitude, on rocky slopes and escarpments. There is no shortage of resource but the trees are scattered through the mountains which stretch from Erigavo in the east to the Ethiopian border in the west. While there are whole areas that have never been tapped, exploiting them will take time.

And what of the resources in Puntland? Guelle explains to me that nobody really knows what is happening there, how the gum is being harvested, by whom, and under what conditions. Puntland, much like the officially recognized Somalia, has descended into lawlessness. In any event, it has become impossible to access the region as a result of travel restrictions and a very real danger. In order to see the resins from Puntland, one has to be content with a visit to the exporters' warehouses in Dubai, where the entire trade in resins converges. There is no talk of Somaliland, Puntland or Yemen to be heard in Dubai, only of business, stock and offers being made to middlemen and brokers. Opacity and secrecy remain the order of the day.

In the late afternoon, Zahra talks me through her different grades of myrrh and frankincense. Their scent fills the room, fanned by the breeze that is picking up. We're drinking tea, enjoying one of those moments of intense pleasure where the fragrance is everywhere but elusive, bewitching yet hidden, drifting here and there on currents of air as night falls. While always warm and balsamic, dark and sensual, the myrrh collected in summer has quite a distinct scent from that collected in winter. The frankincense is aromatic, a blend of terebinth and citrus notes, complex and warm, with an evocation of smokiness even before it is burned. Some frankincense comes from northern Yemen, near the Omani border, far from any war zone. Traditionally, the harvest is carried out by collectors from Somaliland, including some of Guelle's people. I have a better understanding now of the connections between the two shores of the Gulf of Aden as if, over the millennia of this shared proximity, the dhows have created their own wake of fragrance between Africa and Arabia that will linger forever.

---

In a convoy of two other vehicles, each carrying an armed guard, we set off north from Hargeisa on the hunt for myrrh, following 150 kilometers of dirt road. While officially a safety measure, the requirement that one must be accompanied is also in effect a tax and a small source of income for veterans of the war of independence. My pair of armed guards, who are frankly quite friendly, cut a fine figure in their army fatigues. Over the course of the days we spend together, they talk about their years in the resistance, when they first sought cover on the other side of the Ethiopian border, before going on the offensive against the Somali troops from Mogadishu and ultimately emerging victorious. One of them still has a shard of metal in his skull from the war.

As soon as we leave the suburbs of Hargeisa, the desert begins, shifting from sand to an endless rocky landscape. The countryside is studded with several different species of acacia, none of them very tall, all with their typical flattish crown of branches. It is the end of the dry season, the first rains have already fallen on this desert kingdom in which long sharp thorns adorn every tree. The water has brought out tiny green leaves. The acacia seeds—hard, white, pointy pods that can easily be mistaken for flowers—shine in the sun like millions of little daggers. I feel like I have landed in another universe.

Beneath the trees is the yellow ocher of the sandy desert. Far from anywhere, young shepherds are watching their flocks of camels, who know how to nibble the young leaf shoots among the thorns. We see goats, too, and handsome Somali sheep, white with black heads. As we near our destination we come across a village built around a well. The tent-like structures here are decorated more with woven blankets than the plastic that was evident in town. Crowns of branches are laid out on the ground around each house, a thorny barrier that forms an impressive defense against wild animals and possible intruders.

The track has turned from sand to black stone; a lunar landscape but one in which trees continue to grow. The trail climbs, soon we have gained serious height, and at the pass we tip into a striking view of curtains of ocher, red and purple mountains. Below us, traces of green mark the path of a wadi, one of the rivers that are no more than a vast bed of rocks before the rains return to replenish them. As we head down, we cross through a thick forest where the leaves are yet to emerge, frightening away herds of gazelle and antelope.

During the hours we spend in the car, Guelle and Zahra explain how all of Somali society is organized into clans, a structure that has existed for thousands of years. Which clan and sub-clan you belong to determines fundamental aspects of life. You work with your clan, for your clan. This network of relationships, invisible to any outsider, governs everything, from elections to the division of areas for the collection of resin, and "ownership" of each tree.

In the village of Damal, Guelle and his team have set up a new myrrh collection center. Over the last few kilometers, myrrh trees have started to appear. A small tree, barely three meters tall, with a complex system of thorny branches, it is fast growing, its wood soft and not very resistant. The tree itself lives barely longer than forty years. When we arrive, the villagers flock to this white-skinned visitor and the soldiers accompanying him. The young women remain modestly at the entrance to their homes. All the men are keen to show me the myrrh trees that they have just learned how to tap.

The winter season's collection has only just finished, and we walk over to a group of trees that have been selected to demonstrate the tapping procedure. The tapper pulls off pieces of bark a couple of centimeters wide from various places he has chosen on the trunk and main branches. He is using a tool called a *mangaf*, a simple wooden sleeve to protect the hand from thorns that is fitted with two blades. It does not take long for the gum to emerge from the wound and the tapper will come back to the tree in two weeks' time to collect the secretion. The procedure is repeated over the four summer months, which is the most productive season, and is carried out again when winter starts, with the tree allowed to recover between March and June.

Guelle has had a small shop built to stock the first collections from the village, one of thirty centers he has set up over the last couple of years. Bags of fresh myrrh are lined up against the rammed-earth wall of the room, well protected from the light. The gum is soft and shiny, an almost reddish brown, sticky. It gives off a powerful scent in the confined space. Enveloped by fragrance as I lean over to look into the bags, my thoughts of course turn to the three Magi as I try to remember who of Melchior, Gaspar and Balthazar brought with him myrrh to Bethlehem. As it dries out over the next few weeks, it will harden, darken to brown and form clusters which will melt down completely if exposed to heat. Touching the gum with my fingers, I realize why ancient texts note that myrrh is "oily" and should be carried in leather pouches.

After seeing this desert landscape of myrrh trees, we must now head off in search of the frankincense. The night before our departure, I see Guelle growing agitated on the telephone: new police orders have just been issued that prohibit us from traveling east to the main frankincense region we were supposed to be visiting. Guelle had submitted our intended movements well in advance and had received the necessary permits, but, as Zahra explains to me, unforeseen developments like this happen all the time. Guelle springs into action, working his networks. Incomprehension, endless discussions, ministerial meetings. Time is running out. I feel gloomy at the thought that, so close to my goal, I may have to give up on the idea of seeing the frankincense trees; it is like a dream dissipating before my eyes. At the last moment, in response to a suggestion from Zahra, the authorities agree to allow us to visit another area and so it is to the west instead that we set off in search of our frankincense, along a dirt road heading toward Zeila and Djibouti. On hearing this news, my relief is coupled with fresh enthusiasm.

Together with Zahra and Guelle, I look at the maps for the journey that awaits us the following day. I compare them with maps from my own book. There is no question: Zahra confirms that we will be following almost precisely the ancient caravan routes from Harar to Zeila in search of our trees. When I hear this response from our own Queen of Sheba, I have to suppress an emotion which I already know will accompany me on our trip. I tell them about Rimbaud the Ethiopian, the "man with soles of wind," the fanatical walker, speaker of every language, dealer in frankincense, desperately seeking to reinvent himself after abandoning writing at the age of twenty. His poem "Le Bateau ivre" had kept me company as I journeyed down the rivers of the Amazon. Could it now be that we would be walking in the footprints left by the caravaneer of Harar?

A short stretch of sealed road as we leave Hargeisa, then very quickly we find ourselves on a dirt track, flanking the Ethiopian border. We're still in a landscape of acacias, tent villages, shepherds and camels. We are due to meet a farmer from Guelle's network of collectors who is supposed to be taking us to see his trees. After five hours of dirt road, all of a sudden he is there, on the side of the track, having seemingly appeared out of nowhere. He is very thin, wears a worn T-shirt and flip-flops, and carries a plastic bag in his hand. He flashes a radiant smile, climbs into our car and we head off once more. Endless discussions, hesitations, detours and now we are making our way up a dry river bed. Slowly it becomes apparent that we are going around in circles. Guelle realizes from the way our tapper-farmer is looking around in every direction that he has lost his bearings and is too afraid to say so. At last he asks to get out; he will direct us on foot. Once he is walking again, he appears to revive and starts running ahead of the car, gesturing broadly to indicate that he has found the way once more.

The river bed narrows, progress is increasingly difficult. We're advancing between two cliffs over a bed of rocks that are growing ever bigger. In the end, we abandon the car and follow him on foot. Where are the trees and how much longer before we reach them? More gestures and a broad smile are the only response. To a Somali farmer, the concepts of time and distance have quite different meanings. We are heading to the trees. That is all there is to know.

After an hour walking through the rocks, we reach a steep, craggy slope. Zahra is tired and does not want to continue; she no longer believes in this farmer's supposed trees. One of the guards remains with her. As for our guide, he continues at a run, pointing to the top of the hill. Skipping in delight, he leads us onward—me, Guelle and the other guard who is still carrying his weapon—in what will prove to be a long ascent. Every tree we encounter on the way up results in dashed hopes. Guelle confirms that these are still not the *Boswellia* trees, and then suddenly I spot some small trees clinging to the rock face, next to which our guide is gesturing wildly. These are the ones! At last. Young frankincense trees, almost inaccessible all the way up here. Our ascent becomes a scrabble to join him on a narrow, rocky terrace. Out of breath, my heart thumping, I finally cling to the tree I have waited so long to see.

My tree looks like no other I know. The soft wood of its grey trunk is protected by a thin layer of bark. With a generous branch structure, it looks like a young tree, the base of its trunk disappearing between two slabs of rock, as if burying its roots into the heart of the mountain. These trees love the rocks. Far below in the valley is the river bed. The view is staggering, savage, my gaze sweeps the landscape and I feel overcome by vertigo. I realize that before me is indeed the path taken by Rimbaud some one hundred and thirty years earlier; snatches from his book tumble through my mind like an avalanche and I grab hold of the grey trunk. It is March, and the branches are bare of leaves, the wind is picking up. Our tapper has come over to me with Guelle and is explaining that such a young tree, perhaps only ten years old, has not yet been tapped. He takes out his *mangaf*, places it in my hand and shows me where to make the first notch. The wind has risen and it is cold now in the waning sunlight. My hand is trembling. I peel back a piece of bark. Little beads of whitish milk instantly start to pearl on the bare wood. We are standing so close to the trunk that the smell of the nascent frankincense is like a slap in the face, potent and already so recognizable. Our guide points to the cliff opposite and the openings in the rock face. He will store his frankincense harvest in those caves for weeks before heading back down to his farm with a donkey to carry his load. How long has he been tapping these trees? Guelle translates my question. The man is silent and then says at last, still smiling, "We have been doing this forever." Frankincense constantly brings us face to face with the passage of time. The countless years over which it has been collected and traded, the uncounted hours spent roaming the mountains of Somaliland.

The sun is setting, and I experience a rare moment, one I have been waiting for a very long time. My mind flits away. These are the tracks, then, that have been left by that wandering spirit so in-tent on becoming a successful merchant caravaneer that he paid a price of ten years of unmitigated risk and suffering. A despairing and vain quest for some alternative meaning to life, in some other world. April 1891, and the 37-year-old Rimbaud, who is terribly unwell, departs Harar hoping to be nursed in France. Unable to walk on account of a cancerous leg, he is carried for ten days by sixteen porters as far as Zeila. It is his final caravan. Only months later he will die. Clinging to my tree, I scan the landscape. I can make it out now, clearly, a convoy advancing slowly along the track beneath the frankincense trees. I watch it pass, Rimbaud's verse unfurling in my mind. From Queen Hatshepsut to Arthur Rimbaud, the age of frankincense continues. For the tapper at my side, it is an era that has always existed. Trembling, I allow the convoy to slip away into the heady fragrance of those little white pearls. Here, nothing has changed. Not the sky, not the stones, not the paths, nor the rocks where the trees take root, trees from which resin continues to flow.

Time stands still.

# EPILOGUE

## *Travels in alchemy*

Happily, there is no end to these fragrant journeys. Orange blossom, May rose, ylang-ylang, cedar, styrax, iris, guaiac . . . So many other stories which already I am regretting not having shared. I never cease to be impressed by the profusion of paths branching off this perfumed trail, linking up the origins of these fragrances from the deserts of Africa to the depths of the forests in Asia and the Americas, from the shores of the Mediterranean to the tropical regions of the world. I never tire of my encounters and collaboration with the people behind these perfumes: those picking flowers in Bulgaria or India, boiling gum in Andalusia, tapping trees in El Salvador or Laos, those growing patchouli or bergamot, planting sandalwood, distilling vetiver or lavender. So many tasks, some simple, others complex, some archaic, others modern, so many indispensable parts to the process of creating a perfume. Wherever I travel, I am intrigued by the tacit handing down of traditions and knowledge within communities and families, from person to person. Somehow it all feels like a bewitching miracle, that these resins and balsams are still being collected, the farmers' fierce determination, their obsession with ensuring that fragrances smelled for centuries continue to waft through our world.

As they consider what the future may hold for their businesses and for their know-how, producers are wondering if, in tomorrow's world, they will be able to keep producing these scents that have existed forever. What will become of them, of these raw materials, when the world's landscapes are changing at a dizzying pace? The traditional way of life in rural communities is entering a time of turbulence. Faced with the vagaries of climate change, with deforestation and soil depletion, farmers are turning to their screens for images of a different life in the brightly lit cities. For millions of farmers and their children, the attraction of a world perceived as less harsh, one offering more promise, is irresistible. Will any of the farms, the trees, the distilleries, the various jobs that make up the perfume industry be sufficiently appealing and remunerative to retain them? For perhaps the first time since frankincense was collected in ancient times, we can no longer assume the continued existence of perfumes created from natural ingredients. The increase in industry regulation is adding to this uncertainty. There has been a spectacular rush to identify allergens in essential oils, the use of many essences in compositions has been restricted, and naturals—traditionally so alluring and seductive—must now convince people they are safe at every step in the process. Should tomorrow's perfumers be preparing for a diminution in the available natural palette?

Paradoxically, interest on the part of consumers in the Western world in ingredients from the "world of nature" has never been so strong. We all want natural extracts to be used in cosmetics and flavorings for the sake of our health, in aromatherapy for our well-being, and in fragrances for depth and authenticity. At the same time, we are demanding more information and more transparency as to the source of these extracts, the environmental impact of producing them, and the ethics of relationships with farming communities. Fragrance brands and their creators face an unprecedented conundrum, as recent as it is challenging. The increase in demand for natural products is being matched by tighter restrictions governing their use. Sourcing them remains a fragile and complex business, while international standards clamor for ever-improved practices.

The perfume industry is rallying to respond to public expectation. It is implementing charters that outline principles of "ethical sourcing," traceability and community support projects, (and which guarantee quality assurance) to ensure quality and respectability. While the topic is complex and the task challenging, many encouraging projects have seen the light of day in recent years: the building of wells, schools, and clinics that offer emergency care, along with agricultural training centers that provide adolescents with the skills to make a decent living in their villages. Technology in all its forms is making its presence felt in a traditional world: water-efficient agricultural practices, restrictions on the use of pesticides and fertilizers, storage of data on farmers' smartphones. Communities that have for so long been ignored or scorned are resuming their rightful place as the focus of attention. The real revolution underway is in all likelihood the collaboration on the ground between perfume's four major players: farmers, distillers, creators and brands. Perfumery's age-old tradition of secrecy is starting to make way for transparency and ethical responsibility. Finally.

I have long been convinced that the future of natural fragrances lies in the hands of my producer friends: Filip, Gianfranco, Raja, Gigi, Francis, Elisa, Zahra and all the other formidable artisans in this industry. I have loved collaborating with them to discover and develop practices that will encourage them and their children to remain in the industry. The value ascribed by perfumery to frankincense, benzoin and rose oil will be the determining factor in their future. Are we ready to grant the extracts of natural raw materials the status of luxury products that they deserve, a privilege until now reserved for the perfumes in which they are used?

As I draw to the end of my tales, I feel profoundly aware of the unique, almost magical character of these fragrances. My wanderings over so many years have allowed me to witness the alchemy by which aromatic substances produced by the earth are first captured, then refined, before being allowed to disperse in the air. Just as a musician's instrument turns breath into music, a still condenses steam into perfume in a similarly magical sleight of hand. Breath and steam escape their brass and copper, becoming music or fragrance, envoys of the world's beauty.

Millions of flowers, branches, pieces of bark and drops of resin are tipped into stills, all of them bearing the particular mark of their natural environment. They are then distilled and blended, ending up in a flacon as highly concentrated elixirs. And when the bottle is opened and the perfume bursts out, the intricate stories with which it has been entrusted gradually escape back into the world. Hovering briefly over our skin, it wends its way onward, leaving an intimate yet potent trail to linger for hours. Then softly it moves on, whispering the earth's secrets into the breeze, back to the source of the world's scents.

# ACKNOWLEDGMENTS

To my friend, Laurent, who made sure I wrote this book.

To Garance and Victor, in memory of your mother and of Andalusia, thank you for inspiring me to tell these tales.

Éliane, these chronicles bear the hallmark of your many talents: your lucid eye, your robust ideas, your careful rereading. I cannot thank you enough.

To Marie-Hélène, a valued guide through the initial fog, and to Aurélie, who is familiar with these shores, and who is a tireless reader and no-nonsense adviser.

To Xavier, great perfume expert and steadfast friend, and to Pierre, an inexhaustible accomplice and gifted gallivanter.

An emotional thank you to the protagonists in this book for sharing these fragments of their lives and a warm greeting to all my friends at Naturals Together who appear within these pages.

Thank you to Damien Schvartz for sharing his manuscript with me.

Thank you to every perfumer whom it has been my great pleasure to meet, with a special nod to those who appear in this book: Fabrice Pellegrin, Jacques Cavallier, Olivier Cresp, Harry Frémont, Marie Salamagne and Alberto Morillas.

My thanks to Dominique Coutière and Jean-Noël Maisondieu at Biolandes who opened the doors to the profession for me.

To Patrick Firmenich, Armand de Villoutreys, Boet Brinkgreve and Gilbert Ghostine, thank you for Firmenich's trust, for your friendship and for your precious support on our journey back to the source.

Affectionate thoughts to my traveling companions, both current and from yesteryear. And to the wonderful friendships made during my years in Landes: Benoît de Le Sen, José Carlos and Susana, Philippe in Morocco, Siamak, Vessela and all the others.

To Yannick in Grasse, Émilie in Perth, to Jordi and Gemma, and to Bernard from Livelihoods and to my friends at I.F.E.A.T.

To my associates in India, Marc and Sarah, and to Benoît, my comrade in all things vanilla. To the invaluable Helen and Lu Yan in China.

To Julien, Anael and Bastien, the talented three Magi, bearers of images.

# WORDS OF THE TRADE

## *Materials*

EXUDATE: any substance flowing from a wound caused to a tree.

BALSAM: odorous liquid exudate.

GUM/RESIN: exudate from a tree that hardens when exposed to the air, gums being soluble in water, resins being soluble in alcohol.

TEARS: solidified pieces of gum or resin.

ESSENTIAL OIL/ESSENCE: the product obtained from a plant by steam distillation. Being insoluble in water and lighter than water, essential oils/essences are separated from water by a process of decantation.

CONCRETE: odorous wax obtained from a plant by solvent extraction.

ABSOLUTE: that part of the concrete which is soluble in alcohol and able to be used by perfumers.

EXTRACT/RESINOID: product obtained from a plant by alcohol extraction.

FLORAL WATER/ROSEWATER: fragrant water obtained by the distillation of flowers in water.

## *Techniques*

TAPPING: the incision of a tree to make it produce balsams, gums or resins.

ENFLEURAGE: age-old technique to capture the fragrance of flowers by spreading them out on a layer of fat.

DISTILLATION: production of essential oil by circulating steam around plants or a mixture of water and plants.

EXTRACTION: production of concretes/absolutes by circulating solvent around vegetable matter.

STILL OR ALEMBIC: a piece of equipment used to produce floral water or essential oil.

CONDENSATE: a mixture of water and essential oil once the steam from the distillation is condensed back to a liquid state.

FLORENTINE FLASK: a vase designed to collect the separated essential oil that accumulates in a layer on top of the condensed water.

EXTRACTOR: a piece of equipment used to produce concretes or resinoids.

# A NOTE ON THE COVER

We're a collaborative studio, working together on artwork that most often pulls its inspiration from the natural world, interpreting flora and fauna into a visual style that spans from flat and graphic to volumetric and realistic. When *In Search of Perfumes* was brought to us, it was obviously a perfect connection. The bridge this book builds between culture, travel, and nature and its elegant human uses had us immediately inspired.

Our process began by researching the ingredients and cultures using both the book and outside sources. We studied the shapes, forms, and structures of the various plants until we could draw new versions of them to define the compositions and moods we had in mind while retaining their unique visual properties. We got to know the petals of the cistus and their intricate patterns of folds formed by the Andalusian winds. From there it's not a large leap to see the mood we should strike was one balancing beauty, strength, drama, and fragility. We learned that patchouli has a form, and it's one that could supply the perfect arced framing we needed to give compositional structure to the tangle of ingredients as they wrap around the covers and spine.

Thumbnail sketches are small, quick visual ideas, and we used them to try out lots of these different approaches. Some thoughts never pass that phase and are left on the cutting-room floor. Those that were ready were worked into tighter sketches. These were then shared with Stephen Brayda, our art director at HarperVia. He shares them around, there's a dialogue, and the sketches evolve until one feels right to everyone.

To speak toward material process, our sketches are rendered digitally in Procreate and Photoshop, and they move to a final painting in those same applications. We had worked in traditional materials for decades before transitioning to digital media. Our hands and minds interact with our digital tablets in a way that is informed by this background both intuitively and consciously, which is to say, our computers feel like paint and ink to us.

With the idea, subject, mood, and composition all set in the sketches, this makes our final painting process about bringing it all to life through a delicate balance of light, texture, color, and detail. In the end, it feels amazing to bring something into the world that both complements the text it accompanies and, hopefully, can stand on its own as a thing of beauty and emotion.

—Gina Triplett and Matt Curtius
www.ginaandmatt.studio

# INDEX

Here ends Dominique Roques's
*In Search of Perfumes.*

The first edition of this book was printed and
bound at Lakeside Book Company
in Harrisonburg, Virginia, April 2023.

## A NOTE ON THE TYPE

The text of this book was set in Haarlemmer MT Pro, a typeface designed by Frank E. Blokland, founder of Digital Type Library (DTL), the first digital type foundry in the Netherlands. His design was the recreation of a never-produced face by the renowned typographer Jan van Krimpen (1892–1958). The original Haarlemmer was drawn in 1938 to set a new edition of the Bible using Monotype metal typesetting but had to be abandoned when Holland was invaded in World War II. Blokland's homonymous typeface matched Van Krimpen's original concept as nearly as possible, without the technical limitations of metal typesetting, and is widely praised for its elegance and versatility.

HARPERVIA

An imprint dedicated to publishing international voices,
offering readers a chance to encounter other lives and other
points of view via the language of the imagination.